Lecture Notes in Mathematics

Volume 2265

This series reports on new developments in all areas of mathematics and their applications - quickly, informally and at a high level. Mathematical texts analysing new developments in modelling and numerical simulation are welcome. The type of material considered for publication includes:

1. Research monographs
2. Lectures on a new field or presentations of a new angle in a classical field
3. Summer schools and intensive courses on topics of current research.

Texts which are out of print but still in demand may also be considered if they fall within these categories. The timeliness of a manuscript is sometimes more important than its form, which may be preliminary or tentative.

More information about this series at http://www.springer.com/series/304

Jun Kigami

Geometry and Analysis of Metric Spaces via Weighted Partitions

 Springer

Jun Kigami
Graduate School of Informatics
Kyoto University
Kyoto, Japan

ISSN 0075-8434 ISSN 1617-9692 (electronic)
Lecture Notes in Mathematics
ISBN 978-3-030-54153-8 ISBN 978-3-030-54154-5 (eBook)
https://doi.org/10.1007/978-3-030-54154-5

Mathematics Subject Classification: 30F45, 53C23

This Springer imprint is published by the registered company Springer Nature Switzerland AG.
The registered company address is: Gewerbestrasse 11, 6330 Cham, Switzerland

Preface

This monograph reflects what I have been pursuing for the last 10 years or more. The main questions are *what is the natural way to comprehend the "structure" of a space?* and *how can one develop "analysis" on a space from its "structure"?* These are, of course, vague questions since the meanings of the words "structure" and "analysis" are not quite clear. The former can be replaced by "geometry" but this does not make things better. One of the more precise formulations is, for a given space, *what is a natural counterpart of the Euclidean metric on $\mathbb{R}^n$?* Needless to say, rich analysis on $\mathbb{R}^n$ has been developed under the Euclidean metric. A prototype is a class of self-similar sets, including the Sierpinski gasket and carpet, where there is a natural notion of partitions associated with trees and on which Brownian motion has been constructed and studied. This monograph is just a tiny step toward this vast frontier of geometry and analysis of metric spaces but, hopefully, it is a right one at least.

The origin of the story is the notion of successive divisions of compact metric spaces, which appear in many different areas of mathematics such as the construction of self-similar sets, Markov partitions associated with hyperbolic dynamical systems, and dyadic cubes associated with a doubling metric space. The common feature in these is to divide a space into a finite number of subsets, then divide each subset into finitely many pieces, repeating this process again and again. In this monograph, we generalize such successive divisions and call them *partitions*. Given a partition, we consider the notion of a weight function assigning a "size" to each piece of the partition. Intuitively, we believe that a partition and a weight function should provide a "geometry" and an "analysis" on the space of interest. We pursue this idea in Chaps. 2–4. In Chap. 2, the metrizability of a weight function, i.e. the existence of a metric "adapted to" a given weight function, is shown to be equivalent to the Gromov hyperbolicity of the graph associated with the weight function. In Chap. 3, notions such as bi-Lipschitz equivalence, Ahlfors regularity, the volume doubling property, and quasisymmetry are shown to be equivalent to certain properties of weight functions. In particular, we find that quasisymmetry and the volume doubling property are the same notion in the world of weight functions. In Chap. 4, a characterization of the Ahlfors regular conformal dimension of a compact metric space is given as the critical index p of p-energies associated with the partition and the weight function corresponding to the metric.

I express my gratitude to the many colleagues who have given me insights into these subjects. In particular, I would like to thank Professors M. Bonk and B. Kleiner for their suggestions on the directions in which to proceed, the anonymous referees for their valuable comments on the original manuscript, and the Springer editorial and production teams for their sincere efforts to make this monograph much better than its original form.

Last but not least, in this crazy moment of our history, I sincerely wish good health and safety to all, and that I will soon be able to share thoughts face to face with my friends around the world.

Kyoto, Japan   Jun Kigami
August 23, 2020

Contents

Chapter 1
Introduction and a Showcase

1.1 Introduction

Successive divisions of a space have played important roles in many areas of mathematics. One of the simplest examples is the binary division of the unit interval $[0, 1]$ shown in Fig. 1.1. Let $K_\phi = [0, 1]$ and divide K_ϕ in half as $K_0 = [0, \frac{1}{2}]$ and $K_1 = [\frac{1}{2}, 1]$. Next, K_0 and K_1 are divided in half again and yield K_{ij} for each $(i, j) \in \{0, 1\}^2$. Repeating this procedure, we obtain $\{K_{i_1 \ldots i_m}\}_{i_1, \ldots, i_m \in \{0,1\}}$ satisfying

$$K_{i_1 \ldots i_m} = K_{i_1 \ldots i_m 0} \cup K_{i_1 \ldots i_m 1} \tag{1.1.1}$$

for any $m \geq 0$ and $i_1 \ldots i_m \in \{0, 1\}^m$. In this example, there are two notable properties.

The first one is the role of the (infinite) binary tree

$$T_b = \{\phi, 0, 1, 00, 01, 10, 11, 000, 001, 010, 011, \ldots\} = \bigcup_{m \geq 0} \{0, 1\}^m,$$

where $\{0, 1\}^0 = \{\phi\}$. The vertex ϕ is called the root or the reference point and T_b is called the tree with the root (or the reference point) ϕ. Note that the correspondence $i_1 \ldots i_m \to K_{i_1 \ldots i_m}$ determines a map from the binary tree to the collection of compact subsets of $[0, 1]$ with the property (1.1.1).

Secondly, note that $K_{i_1} \supseteq K_{i_1 i_2} \supseteq K_{i_1 i_2 i_3} \supseteq \ldots$ and

$$\bigcap_{m \geq 1} K_{i_1 \ldots i_m} \text{ is a single point} \tag{1.1.2}$$

© The Editor(s) (if applicable) and The Author(s), under exclusive license to Springer Nature Switzerland AG 2020
J. Kigami, *Geometry and Analysis of Metric Spaces via Weighted Partitions*, Lecture Notes in Mathematics 2265, https://doi.org/10.1007/978-3-030-54154-5_1

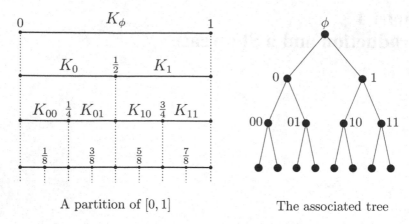

A partition of $[0, 1]$ The associated tree

Fig. 1.1 A partition of the unit interval $[0, 1]$ and the associated tree

for any infinite sequence $i_1 i_2 \ldots$. (Of course, this is the binary expansion and hence the single point is $\sum_{m \geq 1} \frac{i_m}{2^m}$.) In other words, there is a natural map $\sigma : \{0, 1\}^{\mathbb{N}} \to [0, 1]$ given by

$$\sigma(i_1 i_2 \ldots) = \bigcap_{m \geq 1} K_{i_1 \ldots i_m}.$$

Such successive divisions of a compact metric space, which may not be as simple as the above example, appear in various areas in mathematics. One of typical examples is a self-similar set in fractal geometry. A self-similar set is a union of finite contracted copies of itself. Then each contracted copy is again a union of contracted copies and so forth. Another example is the Markov partition associated with a hyperbolic dynamical system. See [1] for details. Also, the division of a metric measure space having the volume doubling property by dyadic cubes can be thought of as another example of such divisions of a space. See Christ[14] for example.

In general, let X be a compact metrizable topological space with no isolated point. The common properties of the above examples are;

(i) There exists a tree T (i.e. a connected graph without loops) with a root ϕ.
(ii) For any vertex p of T, there is a corresponding nonempty compact subset of X denoted by X_p and $X = X_\phi$.
(iii) Every vertex p of T except ϕ has unique predecessor $\pi(p) \in T$ and

$$X_q = \bigcup_{p \in \{p' | \pi(p') = q\}} X_p. \tag{1.1.3}$$

(iv) The totality of edges of T is $\{(\pi(q), q) | q \in T \setminus \{\phi\}\}$.

(v) For any infinite sequence $(p_0, p_1, p_2, \ldots)$ of vertices of X satisfying $p_0 = \phi$ and $\pi(p_{i+1}) = p_i$ for any $i \geq 1$,

$$\bigcap_{i \geq 1} X_{p_i} \text{ is a single point.} \tag{1.1.4}$$

See Fig. 1.2 for an illustration of the idea. Note that the properties (1.1.3) and (1.1.4) correspond to (1.1.1) and (1.1.2) respectively. In this monograph such $\{X_p\}_{p \in T}$ is called a partition of X parametrized by the tree T. (We will give the precise definition in Sect. 2.2.) In addition to the "vertical" edges, which are the edges of the tree, we provide "horizontal" edges to T to describe the combinatorial structure reflecting the topology of X as is seen in Fig. 1.2. More precisely, a pair $(p, q) \in T \times T$ is a horizontal edge if p and q have the same distance from the root ϕ and $X_p \cap X_q \neq \emptyset$. We call T with horizontal and vertical edges the resolution of X associated with the partition.

Another key notion is a weight function on the tree T. Note that a metric and a measure give weights of the subsets of X. More precisely, let d be a metric on X inducing the original topology of X and let μ be a Radon measure on X where $\mu(X_p) > 0$ for any $p \in T$. Define $\rho_d : T \to (0, 1]$ and $\rho_\mu : T \to (0, 1]$ by

$$\rho_d(p) = \frac{\mathrm{diam}(X_p, d)}{\mathrm{diam}(X, d)} \quad \text{and} \quad \rho_\mu(p) = \frac{\mu(X_p)}{\mu(X)},$$

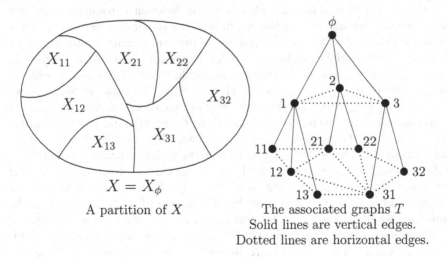

A partition of X

The associated graphs T
Solid lines are vertical edges.
Dotted lines are horizontal edges.

Fig. 1.2 A partition and the associated graphs (up to the 2nd stage)

where $\mathrm{diam}(A, d)$ is the diameter of A with respect to the metric d. Then ρ_d (resp. ρ_μ) is though of as a natural weight of X_p associated with d (resp. μ). In both cases where $\star = d$ or $\star = \mu$, the function $\rho_\star : T \to (0, \infty)$ satisfies

$$\rho_\star(\pi(p)) \geq \rho_\star(p) \tag{1.1.5}$$

for any $p \in T \backslash \{\phi\}$ and

$$\lim_{i \to \infty} \rho_\star(p_i) = 0 \tag{1.1.6}$$

if $\pi(p_{i+1}) = p_i$ for any $i \geq 1$. (To have the second property (1.1.6) in case of $\star = \mu$, we must assume that the measure μ is non-atomic, i.e. $\mu(\{x\}) = 0$ for any $x \in X$.)

As we have seen above, given a metric or a measure, we have obtained a weight function $\rho_\star$ satisfying (1.1.5) and (1.1.6). In this monograph, we are interested in the opposite direction. Namely, given a partition of a compact metrizable topological space parametrized by a tree T, we define the notion of weight functions as the collection of functions from T to $(0, 1]$ satisfying the properties (1.1.5) and (1.1.6). Then our main object of interest is the space of weight functions including those derived from metrics and measures. Naively we believe that a partition and a weight function essentially determine a "geometry" and/or an"analysis" of the original set no matter where the weight function comes. It may come from a metric, a measure or else. Keeping this intuition in mind, we are going to develop a basic theory of weight functions in three closely related directions in this monograph.

The first direction is to study when a weight function is naturally associated with a metric. In brief, our conclusion will be that a power of a weight function is naturally associated with a metric if and only if the rearrangement of the resolution T associated the weight function is Gromov hyperbolic. To be more precise, let $\{X_w\}_{w \in T}$ be a partition of a compact metrizable topological space X with no isolated points and let $\rho : T \to (0, 1]$ be a weight function. In Sect. 2.3, we will define $\delta^\rho_M(\cdot, \cdot)$, which is called the visual pre-metric associated with ρ, in the following way: let Λ^ρ_s be the collection of w's in T where the size $\rho(w)$ is almost s. Define a horizontal edge of Λ^ρ_s as $(w, v) \in \Lambda^\rho_s$ with $X_w \cap X_v \neq \emptyset$. For $r \in (0, 1)$, the rearranged resolution $\widetilde{T}^{\rho,r}$ associated with the weight function ρ is defined as the vertices $\cup_{m \geq 0} \Lambda^\rho_{r^m}$ with the vertical edges from the tree structure of T and the horizontal edges of $\Lambda^\rho_{r^m}$. Then the visual pre-metric $\delta^\rho_M(x, y)$, where $M \geq 1$ is a natural number, for $x, y \in X$ is given by the infimum of s where x and y can be connected by an M-consecutive horizontal edges in Λ^ρ_s. We think that a metric d is naturally associated with the weight function ρ if and only if d and δ^ρ_M are bi-Lipschitz equivalent on $X \times X$. More precisely, we are going to use a phrase"d is adapted to ρ" instead of "d is naturally associated with ρ". The notion of visual pre-metric in this monograph is a counterpart of that of visual pre-metric on the boundary of a Gromov hyperbolic metric space, whose detailed account can be seen in [12, 27] and [18] for example. Now the main conclusion of the first part is Theorem 2.5.12 saying that the hyperbolicity of the rearranged resolution $\tilde{T}^{\rho,r}$ is equivalent to the existence of a metric adapted to some power of the weight function. Moreover, if this is the case, the metric adapted to some power of the weight function is shown to be a visual metric in Gromov's sense.

The second direction is to establish relationships of various relations between weight functions, metrics and measures. For example, Ahlfors regularity and the volume property are relations between measures and metrics. For $\alpha > 0$, a measure μ is α-Ahlfors regular with respect to a metric d if and only if there exist $c_1, c_2 > 0$ such that

$$c_1 r^\alpha \le \mu(B_d(x, r)) \le c_2 r^\alpha,$$

where $B_d(x, r) = \{y | y \in X, d(x, y) < r\}$, for any $r \in (0, \text{diam}(X, d)]$ and $x \in X$. See Definition 3.3.3 for the precise definition of the volume doubling property. On the other hand, bi-Lipschitz and quasisymmetry are equivalence relations between two metrics. (The precise definitions of bi-Lipschitz equivalence and quasisymmetry are given in Definitions 3.1.9 and 3.6.1 respectively.) Regarding those relations, we are going to claim the following relationships

$$\text{bi-Lipschitz} = \text{Ahlfors regularity} = \text{being adapted} \tag{1.1.7}$$

and

$$\text{the volume doubling property} = \text{quasisymmetry.} \tag{1.1.8}$$

in the framework of weight functions. To illustrate the first claim more explicitly, let us introduce the notion of bi-Lipschitz equivalence of weight functions. Two weight functions ρ_1 and ρ_2 are said to be bi-Lipschitz equivalent if and only if there exist $c_1, c_2 > 0$ such that

$$c_1 \rho_1(p) \le \rho_2(p) \le c_2 \rho_1(p)$$

for any $p \in T$. Now the first statement (1.1.7) can be resolved into three parts as follows: let ρ_1 and ρ_2 be two weight functions.

Claim 1: Suppose that $\rho_1 = \rho_{d_1}$ and $\rho_2 = \rho_{d_2}$ for metrics d_1 and d_2 on X. Then ρ_1 and ρ_2 are bi-Lipschitz equivalent if and only if d_1 and d_2 are bi-Lipschitz equivalent as metrics.

Claim 2: Suppose that $\rho_1 = \rho_d$ and $\rho_2 = \rho_\mu$ for a metric d on X and a Radon measure μ on X. Then $(\rho_1)^\alpha$ and ρ_2 are bi-Lipschitz equivalent if and only if μ is α-Ahlfors regular with respect to d.

Claim 3: Suppose that $\rho_1 = \rho_d$ for a metric d on X, then ρ_1 and ρ_2 are bi-Lipschitz equivalent if and only if the metric d is adapted to the weight function ρ_2.

One can find the precise statement in Theorem 1.2.11 in the case of partitions of S^2. The second statement (1.1.8) is rationalized in the same manner. See Theorem 1.2.12 for the exact statement in the case of S^2 for example.

The third direction is a characterization of Ahlfors regular conformal dimension. The Ahlfors regular conformal dimension, AR conformal dimension for short, of a metric space (X, d) is defined as

$$\dim_{AR}(X, d) = \inf\{\alpha | \text{there exist a metric } \rho \text{ on } X \text{ and a Borel regular measure } \mu \text{ on}$$
$$X \text{ such that } \rho \underset{QS}{\sim} d \text{ and } \mu \text{ is } \alpha\text{-Ahlfors regular with respect to } \rho\},$$

where "$\rho \underset{QS}{\sim} d$" means that the two metrics ρ and d are quasisymmetric to each other. Note that α is the Hausdorff dimension of (X, d) if there exists a measure μ which is α-Ahlfors regular with respect to d. In [13], Carrasco Piaggio has given a characterization of Ahlfors regular conformal dimension in terms of the critical exponent of p-combinatorial modulus of discrete path families. Given the results from the previous part, we have obtained the ways to express the notions of quasisymmetry and Ahlfors regularity in terms of weight functions. So we are going to translate Carrasco Piaggio's work into our framework. However, we are going to use the critical exponent of p-energy instead of p-combinatorial modulus in our work.[1] Despite the difference between p-energy and p-modulus, we will make essential use of Carrasco Piaggio's ideas, which have been quietly embedded in his paper [13]. Furthermore, we are going to define the notion of p-spectral dimension and present a relation between Ahlfors regular conformal dimension and p-spectral dimension. In particular, for $p = 2$, the 2-spectral dimension has been know to appear in the asymptotic behavior of the Brownian motion and the eigenvalue counting function of the Laplacian on certain fractals like the Sierpinski gasket and the Sierpinski carpet. See [4, 5] and [23] for example. For the Sierpinski carpet, we will show that the 2-spectral dimension gives an upper bound of Ahlfors regular conformal dimension.

One of the main reasons for recent interest in Ahlfors regular conformal dimension is its close connection with the Cannon's conjecture, although we are not going to pursue such a direction in this monograph.

Conjecture (Cannon) Let G be a Gromov hyperbolic group whose boundary $\partial_\infty G$ is homeomorphic to the two dimensional sphere S^2. Let d be a visual metric on $\partial_\infty G$. Then $(\partial_\infty G, d)$ is quasisymmetric to S^2 with the Euclidean metric.

For this conjecture, Bonk and Kleiner have given the following partial answer in [7].

---

[1]This idea of characterizing AR conformal dimension by p-energies was brought to the author by B. Kleiner in a personal communication. In fact, Keith and Kleiner had obtained a result which is comparable with that of Carrasco Piaggio [13]. See the discussions after [10, Proposition 1.5].

Theorem 1.1.1 (Bonk-Kleiner) *Let G be a Gromov hyperbolic group whose boundary $\partial_\infty G$ is homeomorphic to S^2 and let d be a visual metric on $\partial_\infty G$. Assume that Ahlfors regular conformal dimension is attained, i.e. there exist a metric ρ on $\partial_\infty G$ which is quasisymmetric to d and a Borel regular measure μ which is $\dim_{AR}(X, d)$-Ahlfors regular with respect to ρ. Then $(\partial_\infty G, d)$ is quasisymmetric to S^2 with the Euclidean metric.*

So, the last piece to prove Cannon's conjecture is the attainability of Ahlfors regular conformal dimension. Besides Canon's conjecture, the attainability of Ahlfors regular conformal dimension is an interesting problem by itself and not solved except for limited classes of spaces like domains in $\mathbb{R}^n$. In particular, it is still open even in the case of the Sierpinski carpet at this moment.

There have been other works on an analytic characterization of Ahlfors regular conformal dimension. In [11], Bourdon and Pajot have used the critical exponent of ℓ_p-cohomologies $p(X, d)$. In general, $p(X, d) \leq \dim_{AR}(X, d)$ and the equality holds if Ahlfors regular conformal dimension is attainable with a Loewner metric. Lidquist has used the critical exponent of weak capacities Q_w in [26]. He has shown that $Q_w \leq \dim_{AR}(X, d)$ and the equality has been shown to hold if Ahlfors regular conformal dimension is attainable. For certain class of self-similar sets, Shimizu has shown $p_*(G) \leq \dim_{AR}(X, d)$, where $p_*(G)$ is the parabolic index of a blow-up G, which is an infinite graph, of a self-similar set X in [28].

One of the ideas behind this study is to approximate a space by a series of graphs. Such an idea has already been explored in association with hyperbolic geometry. For example, in [15] and [11], they have constructed an infinite graph whose hyperbolic boundary is homeomorphic to given compact metric space. They have constructed a series of coverings of the space, which is a counterpart of our partition, and then built a graph from the series. In [13], Carrasco Piaggio has utilized this series of coverings to study the Ahlfors regular conformal dimension of the space. His notion of "relative radius" essentially corresponds to our weight functions. In our framework, the original space is homeomorphic to the analogue of hyperbolic boundary of the resolution T of X even if it is not hyperbolic in the sense of Gromov. See Theorem 2.5.5 for details. In other words, the resolution T of X is a version of hyperbolic filling of the original space X. (See [9] for the notion of hyperbolic fillings.) In this respect, our study in this monograph may be thought of as a theory of weighted hyperbolic fillings.

The organization of this monograph is as follows. In Sect. 1.2, we give a summary of the main results of this monograph in the case of the 2-dimensional sphere as a showcase of the full theory. In Sect. 2.1, we give basic definitions and notations on trees. Section 2.2 is devoted to the introduction of partitions and related notions. In Sect. 2.3, we define the notion of weight function and the associated "visual pre-metric". We study our first question mentioned above, namely, when a weight function is naturally associated with a (power of) metric in Sect. 2.4. In Sect. 2.5, we are going to relate this question to the hyperbolicity of certain graph, called a rearrangement above, associated with a weight function. Section 3.1 is devoted to justifying the statement (1.1.7). In Sects. 3.2, 3.3, 3.5 and 3.6, we will study the rationalized version of (1.1.8) as a mathematical statement. In particular, in Sect. 3.3, we introduce the key notion of being "gentle". In Sect. 3.4, we apply our general theory to certain class of subsets of the square and obtain concrete (counter) examples. From Sect. 4.1, we will start arguing a characterization of Ahlfors regular conformal dimension. From Sect. 4.1–4.5, we discuss how to obtain a pair of a metric d and a measure μ where μ is α-Ahlfors regular with respect to d for a given order α. The main result of these sections is Theorem 4.5.1. In Sect. 4.6, we will give a characterization of the Ahlfors regular conformal dimension as a critical index p of p-energies. Then we will show the relation of the Ahlfors regular conformal dimension and p-spectral dimension in Sect. 4.7. Additionally, we will give another characterization of the Ahlfors regular conformal dimension by p-modulus of curve families in Sect. 4.8. This recovers the original result by Carrasco Piaggio [13]. Finally in Sect. B, we present the whereabouts of definitions, notations, and conditions appearing in this monograph for reader's sake.

Remark on the Usage of min, max, sup, inf **and** $\sum$
Throughout this monograph, if $A \subseteq [0, \infty)$ is empty, then we set

$$\min A = \max A = \sup A = \inf A = 0.$$

Moreover, if $f : A \to \mathbb{R}$ and $A = \emptyset$, then we set

$$\sum_{x \in A} f(x) = 0.$$

Such situations may happen if our target space (X, d) is not connected.

1.2   Summary of the Main Results; the Case of 2-dim. Sphere

In this section, we summarize our main results in this monograph in the case of a 2-dimensional sphere S^2 (or in other words, the Riemann sphere), which is denoted by X in what follows. We use d_s to denote the standard spherical geodesic metric on X. Set

$$\mathcal{U} = \{A | A \subseteq X, \text{closed}, \text{int}(A) \neq \emptyset, \partial A \text{ is homeomorphic to the circle } S^1.\}$$

First we divide X into finite number of subsets $X_1, \ldots, X_{N_0}$ belonging to $\mathcal{U}$, i.e.

$$X = \bigcup_{i=1}^{N_0} X_i.$$

We assume that $X_i \cap X_j = \partial X_i \cap \partial X_j$ if $i \neq j$. Next each X_i is divided into finite number of its subsets $X_{i1}, X_{i2}, \ldots, X_{iN_i} \in \mathcal{U}$ in the same manner as before. Repeating this process, we obtain $X_{i_1 \ldots i_k}$ for any $i_1 \ldots i_k$ satisfying

$$X_{i_1 \ldots i_k} = \bigcup_{j=1,\ldots,N_{i_1 \ldots i_k}} X_{i_1 \ldots i_k j} \qquad (1.2.1)$$

and if $i_1 \ldots i_k \neq j_1 \ldots j_k$, then

$$X_{i_1 \ldots i_k} \cap X_{j_1 \ldots j_k} = \partial X_{i_1 \ldots i_k} \cap \partial X_{j_1 \ldots j_k}. \qquad (1.2.2)$$

Note that (1.2.1) is a counterpart of (1.1.3). Next define

$$T_k = \{i_1 \ldots i_k | i_j \in \{1, \ldots, N_{i_1 \ldots i_{j-1}}\} \text{ for any } j = 1, \ldots, k-1\}$$

for any $k = 0, 1, \ldots$, where T_0 is a one point set $\{\phi\}$. Let $T = \cup_{k \geq 0} T_k$. Then T is naturally though of as a (non-directed) tree whose edges are given by the totality of $(i_1 \ldots i_k, i_1 \ldots i_k i_{k+1})$. We regard the correspondence $w \in T$ to $X_w \in \mathcal{U}$ as a map from T to $\mathcal{U}$, which is denoted by $\mathcal{X}$. Namely, $\mathcal{X}(w) = X_w$ for any $w \in T$. Note that $\mathcal{X}(\phi) = X$. Define

$$\Sigma = \{i_1 i_2 \ldots | i_1 \ldots i_k \in T_k \text{ for any } k \geq 0\},$$

which is the "boundary" of the infinite tree T.

Furthermore we assume that for any $i_1 i_2 \ldots \in \Sigma$

$$\bigcap_{k=1,2,\ldots} X_{i_1 \ldots i_k}$$

is a single point, which is denoted by $\sigma(i_1 i_2 \ldots)$. Note that σ is a map from Σ to X. This assumption corresponds to (1.1.4) and hence the map $\mathcal{X}$ is a partition of X parametrized by the tree T, i.e. it satisfies the conditions (i), (ii), (iii), (iv) and (v) in the introduction. Since $X = \cup_{w \in T_k} X_w$ for any $k \geq 0$, this map σ is surjective.

In [8, Chapter 5], the authors have constructed "cell decomposition" associated with an expanding Thurston map. This "cell decomposition" is, in fact, an example of a partition formulated above.

Throughout this section, for simplicity, we assume the following conditions (SF) and (TH), where (SF) is called strong finiteness in Definition 2.2.4 and (TH) ensures the thickness of every exponential weight function. See Definition 3.1.19 for the "thickness" of a weight function.

(SF)

$$\#(\sigma^{-1}(x)) < +\infty, \qquad (1.2.3)$$

where $\#(A)$ is the number of elements in a set A.

(TH)   There exists $m \geq 1$ such that for any $w = i_1 \ldots i_n \in T$, there exists $v = i_1 \ldots i_n i_{n+1} \ldots i_{n+m} \in T$ such that $X_v \subseteq \text{int}(X_w)$.

The main purpose of this monograph is to describe metrics and measures of X from a given weight assigned to each piece X_w of the partition $\mathcal{X}$.

Definition 1.2.1 A map $g : T \to (0, 1]$ is called a weight function if and only if it satisfies the following conditions (G1), (G2) and (G3).

(G1) $g(\phi) = 1$
(G2) $g(i_1 \ldots i_k) \geq g(i_1 \ldots i_k i_{k+1})$ for any $i_1 \ldots i_k \in T$ and $i_1 \ldots i_k i_{k+1} \in T$.
(G3)

$$\lim_{k \to \infty} \sup_{w \in T_k} g(w) = 0.$$

Define

$$\mathcal{G}(T) = \{g | g : T \to (0, 1] \text{ is a weight function.}\}$$

Moreover, we define following conditions (SpE) and (SbE), which represent "super-exponential" and "sub-exponential" respectively:

(SpE)   There exists $\lambda \in (0, 1)$ such that

$$g(i_1 \ldots i_k i_{k+1}) \geq \lambda g(i_1 \ldots i_k)$$

for any $i_1 \ldots i_k \in T$ and $i_1 \ldots i_k i_{k+1} \in T$.
(SbE)   There exist $m \in \mathbb{N}$ and $\gamma \in (0, 1)$ such that

$$g(i_1 \ldots i_k i_{k+1} \ldots i_{k+m}) \leq \gamma g(i_1 \ldots i_k)$$

for any $i_1 \ldots i_k \in T$ and $i_1 \ldots i_k i_{k+1} \ldots i_{k+m} \in T$.

Set

$$\mathcal{G}_e(T) = \{g | g : T \to (0, 1] \text{ is a weight function}$$
$$\text{satisfying (SpE) and (SbE)}.\}.$$

Metrics and measures on X naturally have associated weight functions.

Definition 1.2.2 Set

$$\mathcal{D}(X) = \{d | d \text{ is a metric on } X \text{ which produces the original topology of } X,$$
$$\text{and diam}(X, d) = 1\}$$

and

$$\mathcal{M}(X) = \{\mu | \mu \text{ is a Borel regular probability measure on } X, \mu(\{x\}) = 0$$

$$\text{for any } x \in T \text{ and } \mu(O) > 0 \text{ for any non-empty open set } O \subseteq X\}.$$

For any $d \in \mathcal{D}(X)$, define $g_d : T \to (0, 1]$ by $g_d(w) = \text{diam}(X_w, d)$ and for any $\mu \in \mathcal{M}(X)$, define $g_\mu : T \to (0, 1]$ by $g_\mu(w) = \mu(X_w)$ for any $w \in T$.

From Proposition 2.3.5, we have the following fact.

Proposition 1.2.3 *If $d \in \mathcal{D}(X)$ and $\mu \in \mathcal{M}(X)$, then g_d and g_μ are weight functions.*

So a metric $d \in \mathcal{D}(X)$ has associated weight function g_d. How about the converse direction, i.e. for a given weight function g, is there a metric d such that $g = g_d$? To make this question more rigorous and flexible, we define the notion of "visual pre-metric" $\delta_M^g(\cdot, \cdot)$ associated with a weight function g.

Definition 1.2.4 Let $g \in \mathcal{G}(T)$. Define

$$\Lambda_s^g = \{i_1 \dots i_k | i_1 \dots i_k \in T, g(i_1 \dots i_{k-1}) > s \geq g(i_1 \dots i_k)\}$$

for $s \in (0, 1]$ and

$$\delta_M^g(x, y) = \inf\{s | \text{there exist } w(1), \dots, w(M + 1) \in \Lambda_s^g \text{ such that}$$

$$x \in X_{w(1)}, y \in X_{w(M+1)} \text{ and}$$

$$X_{w(j)} \cap X_{w(j+1)} \neq \emptyset \text{ for any } j = 1, \dots, M\}$$

for $M \geq 0$, $x, y \in X$. A weight function is called uniformly finite if and only if

$$\sup_{s \in (0,1], w \in \Lambda_s^g} \#(\{v | v \in \Lambda_s^g, X_w \cap X_v \neq \emptyset\}) < +\infty.$$

Although $\delta_M^g(x, y) \geq 0$, $\delta_M^g(x, y) = 0$ if and only if $x = y$ and $\delta_M^g(x, y) = \delta_M^g(y, x)$, the quantity δ_M^g may not satisfy the triangle inequality in general. The visual pre-metric $\delta_M^g(x, y)$ is a counterpart of the visual metric defined in [8]. See Sect. 2.3 for details.

If the pre-metric $\delta_M^g(\cdot, \cdot)$ is bi-Lipschitz equivalent to a metric d, we consider d as a metric which is naturally associated with the weight function g.

Definition 1.2.5 Let $M \geq 1$

(1) A metric $d \in \mathcal{D}(X)$ is said to be M-adapted to a weight function $g \in \mathcal{G}(X)$ if and only if there exist $c_1, c_2 > 0$ such that

$$c_1 d(x, y) \leq \delta_M^g(x, y) \leq c_2 d(x, y)$$

for any $x, y \in X$.
(2) A metric d is said to be M-adapted if and only if it is M-adapted to g_d and it is said to be adapted if it is M-adapted for some $M \geq 1$.
(3) Define

$$\mathcal{D}_{A,e}(X) = \{d | d \in \mathcal{D}(X), \ g_d \in \mathcal{G}_e(T) \text{ and } d \text{ is adapted}\},$$

$$\mathcal{M}_e(X) = \{\mu | \mu \in \mathcal{M}(X), \ g_\mu \in \mathcal{G}_e(T)\}.$$

The value M really matters. See Example 3.4.9 for an example.

The following definition is used to describe an equivalent condition for the existence of an adapted metric in Theorem 1.2.7.

Definition 1.2.6 Let $g \in \mathcal{G}(T)$. For $r \in (0, 1)$, define $\widetilde{T}^{g,r} = \sqcup_{m \geq 0} \Lambda_{r^m}^g$, where the symbol $\sqcup$ means a disjoint union. Define the horizontal edges $E_{g,r}^h$ and the vertical edges $E_{g,r}^v$ of $\widetilde{T}^{g,r}$ as

$$E_{g,r}^h = \{(w, v) | w, v \in \Lambda_{r^m}^g \text{ for some } m \geq 0, w \neq v, X_w \cap X_v \neq \emptyset\}$$

and

$$E_{g,r}^v = \{(w, v) | w \in \Lambda_{r^m}^g, v \in \Lambda_{r^{m+1}}^g \text{ for some } m \geq 0, X_w \supseteq X_v\}$$

respectively.

The following theorem is a special case of Theorem 2.5.12.

Theorem 1.2.7 *Let $g \in \mathcal{G}(X)$. There exist $M \geq 1$, $\alpha > 0$ and a metric $d \in \mathcal{D}(X)$ such that d is M-adapted to g^α if and only if the graph $(\widetilde{T}^{g,r}, E_{g,r}^h \cup E_{g,r}^v)$ is Gromov hyperbolic for some $r \in (0, 1)$. Moreover, if this is the case, then X is homeomorphic to the hyperbolic boundary of $(\widetilde{T}^{g,r}, E_{g,r}^h \cup E_{g,r}^v)$ and the adapted metric d is a visual metric in the Gromov sense.*

Next, we define two equivalent relations $\underset{BL}{\sim}$ and $\underset{GE}{\sim}$ on the collection of exponential weight functions. Later, we are going to identify these with known relations according to the types of weight functions.

Definition 1.2.8 Let $g, h \in \mathcal{G}_e(T)$.

(1) g and h are said to be bi-Lipschitz equivalent if and only if there exist $c_1, c_2 > 0$ such that

$$c_1 g(w) \leq h(w) \leq c_2 g(w)$$

for any $w \in T$. We write $g \underset{BL}{\sim} h$ if g and h are bi-Lipschitz equivalent.

(2) h is said to be gentle to g if and only if there exists $\gamma > 0$ such that if $w, v \in \Lambda_s^g$ and $X_w \cap X_v \neq \emptyset$, then $h(w) \leq \gamma h(v)$. We write $g \underset{GE}{\sim} h$ if h is gentle to g.

Clearly, $\underset{BL}{\sim}$ is an equivalence relation. On the other hand, the fact that $\underset{GE}{\sim}$ is an equivalence relation is not quite obvious and going to be shown in Theorem 3.5.2.

Proposition 1.2.9 *The relations $\underset{BL}{\sim}$ and $\underset{GE}{\sim}$ are equivalent relations in $\mathcal{G}_e(T)$. Moreover, if $g \underset{BL}{\sim} h$, then $g \underset{GE}{\sim} h$.*

Some of the properties of a weight function are invariant under the equivalence relation $\underset{GE}{\sim}$ as follows.

Proposition 1.2.10

(1) *Being uniformly finite is invariant under the equivalence relation $\underset{GE}{\sim}$, i.e. if $g \in \mathcal{G}_e(T)$ is uniformly finite, $h \in \mathcal{G}_e(T)$ and $g \underset{GE}{\sim} h$, then h is uniformly finite.*

(2) *The hyperbolicity of $\widetilde{T}^{g,r}$ is invariant under the equivalence relation $\underset{GE}{\sim}$.*

The statements (1) and (2) of the above theorem are the special cases of Theorems 3.5.7 and 3.5.9 respectively.

The next theorem shows that bi-Lipschitz equivalence of weight functions can be identified with other properties according to the types of involved weight functions.

Theorem 1.2.11

(1) *For $d, \rho \in \mathcal{D}_{A,e}(X)$, $g_d \underset{BL}{\sim} g_\rho$ if and only if d and g are bi-Lipschitz equivalent as metrics.*

(2) *For $\mu, v \in \mathcal{M}(X)$, $g_\mu \underset{BL}{\sim} g_v$ if and only if there exist $c_1, c_2 > 0$ such that*

$$c_1 \mu(A) \leq v(A) \leq c_2 \mu(A)$$

for any Borel set $A \subseteq X$.

(3) *For $g \in \mathcal{G}_e(X)$ and $d \in \mathcal{D}_{A,e}(X)$, $g \underset{BL}{\sim} g_d$ if and only if d is M-adapted to g for some $M \geq 1$.*

(4) *For $d \in \mathcal{D}_{A,e}(X)$, $\mu \in \mathcal{M}(X)$ and $\alpha > 0$, $(g_d)^\alpha \underset{BL}{\sim} g_\mu$ and g_d is uniformly finite if and only if μ is α-Ahlfors regular with respect to d, i.e. there exist $c_1, c_2 > 0$ such that*

$$c_1 r^\alpha \le \mu(B_d(x, r)) \le c_2 r^\alpha$$

for any $r > 0$ and $x \in X$.

The statements (1), (2), (3) and (4) of the above theorem follow from Corollary 3.1.10, Theorems 3.1.4, 3.1.14 and 3.1.21 respectively.

The gentle equivalence relation is identified with "quasisymmetry" between two metrics and with "volume doubling property" between a metric and a measure.

Theorem 1.2.12

(1) *Let $d \in \mathcal{D}_{A,e}(X)$ and let $\mu \in \mathcal{M}(X)$. Then $g_\mu \in \mathcal{G}_e(T)$, $g_d \underset{GE}{\sim} g_\mu$ and g_d is uniformly finite if and only if μ has the volume doubling property with respect to d, i.e. there exists $C > 0$ such that*

$$\mu(B_d(x, 2r)) \le C\mu(B_d(x, r))$$

for any $r > 0$ and $x \in X$.

(2) *For $d \in \mathcal{D}_{A,e}(X)$ and $\rho \in \mathcal{D}(X)$, d is quasisymmetric to ρ if and only if $\rho \in \mathcal{D}_{A,e}(X)$ and $g_d \underset{GE}{\sim} g_\rho$.*

The statement (1) of the above theorem follows from Proposition 3.3.6 and Theorem 3.3.9-(2). Note that the condition (TH) implies (TH1) appearing in Theorem 3.2.3. Consequently every exponential weight function is thick by Theorem 3.2.3. The statement (2) is immediate from Corollary 3.6.7.

In [8, Section 17], the authors have shown that the visual metric is quasisymmetric to the chordal metric which is bi-Lipschitz equivalent to the standard geodesic metric d_S on S^2 for certain class of expanding Thurston maps. In view of their proof, they have essentially shown a counterpart of the condition given in Theorem 1.2.12-(2).

Next we present a characterization of the Ahlfors regular conformal dimension using the critical index p of p-energies.

Definition 1.2.13 Let $g \in \mathcal{G}_e(T)$ and let $r \in (0, 1)$. For $A \subseteq \Lambda^g_{r^m}$, $w \in \Lambda^g_{r^m}$, $M \ge 1$ and $n \ge 0$, define

$$S^n(A) = \{v | v \in \Lambda^g_{r^{m+n}}, X_v \subseteq \cup_{u \in A} X_u\}$$

and

$$\Gamma_M^g(w) = \{v | v \in \Lambda_{r^m}^g, \text{ there exists } (v(0), v(1), \dots, v(M)) \text{ such that}$$

$$v(0) = w \text{ and } (v(i), v(i+1)) \in E_{g,r}^h \text{ for any } i = 0, \dots, M-1\}.$$

The set $S^n(A)$ corresponds to the refinement of A in $\Lambda_{r^{m+n}}^g$ and the set $\Gamma_M^g(w)$ is the M-neighborhood of w in $\Lambda_{r^m}^g$.

Definition 1.2.14 Let $g \in \mathcal{G}_e(T)$ and let $r \in (0, 1)$. For $p > 0$, $w \in \widetilde{T}^{g,r}$, $M \geq 1$ and $n \geq 0$, define

$$\mathcal{E}_{M,p,w,n}^g = \inf \left\{ \sum_{(u,v) \in E_{g,r}^h, u, v \in \Lambda_{r^{m+n}}^g} |f(u) - f(v)|^p \right|$$

$$f : \Lambda_{r^{m+n}}^g \to \mathbb{R}, \ f|_{S^n(w)} = 1, \ f|_{\Lambda_{r^{m+n}}^g \setminus S^n(\Gamma_M^g(w))} = 0 \right\},$$

where the integer m is chosen so that $w \in \Lambda_{r^m}^g$, and

$$\mathcal{E}_{M,p}^g = \liminf_{m \to \infty} \sup_{w \in \widetilde{T}^{g,r}} \mathcal{E}_{M,p,w,m}^g.$$

By Theorem 4.6.4, we have the following characterization of the Ahlfors regular conformal dimension of (X, d) in terms of $\mathcal{E}_{M,p}^g$.

Theorem 1.2.15 *Let $d \in \mathcal{D}_{A,e}(X)$ and set $g = g_d$. Assume that d is uniformly finite and M-adapted. If $\mathcal{E}_{M,p}^g = 0$, then there exist $\rho \in \mathcal{D}_{A,e}(X)$ and $\mu \in \mathcal{M}_e(X)$ such that μ is p-Ahlfors regular with respect to ρ and ρ is quasisymmetric to d. Moreover, the Ahlfors regular conformal dimension of (X, d) is (finite and) given by $\inf\{p | \mathcal{E}_{M,p}^g = 0\}$.*

Chapter 2
Partitions, Weight Functions and Their Hyperbolicity

2.1 Tree with a Reference Point

In this section, we review basic notions and notations on a tree with a reference point.

Definition 2.1.1 Let T be a countably infinite set and let $\mathcal{A} : T \times T \to \{0, 1\}$ which satisfies $\mathcal{A}(w, v) = \mathcal{A}(v, w)$ and $\mathcal{A}(w, w) = 0$ for any $w, v \in T$. We call the pair $(T, \mathcal{A})$ a (non-directed) graph with the vertices T and the adjacent matrix $\mathcal{A}$. An element $(u, v) \in T \times T$ is called an edge of $(T, \mathcal{A})$ if and only if $\mathcal{A}(u, v) = 1$. We will identify the adjacent matrix $\mathcal{A}$ with the collection of edges $\{(u, v) | u, v \in T, \mathcal{A}(u, v) = 1\}$.

(1) The set $\{v | \mathcal{A}(w, v) = 1\}$ is called the neighborhood of w in $(T, \mathcal{A})$. $(T, \mathcal{A})$ is said to be locally finite if the neighborhood of w is a finite set for any $w \in T$.

(2) For $w_0, \ldots, w_n \in T$, $(w_0, w_1, \ldots, w_n)$ is called a path between w_0 and w_n if $\mathcal{A}(w_i, w_{i+1}) = 1$ for any $i = 0, 1, \ldots n - 1$. A path $(w_0, w_1, \ldots, w_n)$ is called simple if and only if $w_i \neq w_j$ for any i, j with $0 \le i < j \le n$ and $|i - j| < n$.

(3) $(T, \mathcal{A})$ is called a (non-directed) tree if and only if there exists a unique simple path between w and v for any $w, v \in T$ with $w \neq v$. For a tree $(T, \mathcal{A})$, the unique simple path between two vertices w and v is called the geodesic between w and v and denoted by $\overline{wv}$. We write $u \in \overline{wv}$ if $\overline{wv} = (w_0, w_1, \ldots, w_n)$ and $u = w_i$ for some i.

In this monograph, we always fix a point in a tree as the root of the tree and call the point the reference point.

J. Kigami, *Geometry and Analysis of Metric Spaces via Weighted Partitions*,
Lecture Notes in Mathematics 2265, https://doi.org/10.1007/978-3-030-54154-5_2

Definition 2.1.2 Let $(T, \mathcal{A})$ be a tree and let $\phi \in T$. The triple $(T, \mathcal{A}, \phi)$ is called a tree with a reference point ϕ.

(1) Define $\pi : T \to T$ by

$$\pi(w) = \begin{cases} w_{n-1} & \text{if } w \neq \phi \text{ and } \overline{\phi w} = (w_0, w_1, \ldots, w_{n-1}, w_n), \\ \phi & \text{if } w = \phi \end{cases}$$

and set $S(w) = \{v | \mathcal{A}(w, v) = 1\} \backslash \{\pi(w)\}$.

(2) For $w \in T$, we define $|w| = n$ if and only if $\overline{\phi w} = (w_0, w_1, \ldots, w_n)$ with $w_0 = \phi$ and $w_n = w$. Moreover, we set $(T)_m = \{w | w \in T, |w| = m\}$.

(4) An infinite sequence of vertices $(w_0, w_1, \ldots)$ is called an infinite geodesic ray originated from w_0 if and only if $(w_0, \ldots, w_n) = \overline{w_0 w_n}$ for any $n \geq 0$. Two infinite geodesic rays $(w_0, w_1, \ldots)$ and $(v_0, v_1, \ldots)$ are equivalent if and only if there exists $k \in \mathbb{Z}$ such that $w_{n+k} = v_n$ for sufficiently large n. An equivalent class of infinite geodesic rays is called an end of T. We use Σ to denote the collection of ends of T.

(5) Define Σ^w as the collection of infinite geodesic rays originated from $w \in T$. For any $v \in T$, Σ_v^w is defined as the collection of elements of Σ^w passing through v, namely

$$\Sigma_v^w = \{(w, w_1, \ldots) | (w, w_1, \ldots) \in \Sigma^w, w_n = v \text{ for some } n \geq 1\}.$$

Remark Strictly, the notations like π and $|\cdot|$ should be written as $\pi^{(T, \mathcal{A}, \phi)}$ and $|\cdot|_{(T, \mathcal{A}, \phi)}$ respectively. In fact, if we will need to specify the tree in question, we are going to use such explicit notations.

One of the typical examples of a tree is the infinite binary tree. In the next example, we present a class of trees where $\#(S(w))$ is independent of $w \in T$.

Example 2.1.3 Let $N \geq 2$ be an integer. Let $T_m^{(N)} = \{1, \ldots, N\}^m$ for $m \geq 0$. (We let $T_0^{(N)} = \{\phi\}$, where ϕ represents an empty sequence.) We customarily write $(i_1, \ldots, i_m) \in T_m^{(N)}$ as $i_1 \ldots i_m$. Set $T^{(N)} = \cup_{m \geq 0} T_m^{(N)}$. Define $\pi : T^{(N)} \to T^{(N)}$ by $\pi(i_1 \ldots i_m i_{m+1}) = i_1 \ldots i_m$ for $m \geq 0$ and $\pi(\phi) = \phi$. Furthermore, define

$$\mathcal{A}_{wv}^{(N)} = \begin{cases} 1 & \text{if } w \neq v, \text{ and either } \pi(w) = v \text{ or } \pi(v) = w, \\ 0 & \text{otherwise.} \end{cases}$$

Then $(T^{(N)}, \mathcal{A}^{(N)}, \phi)$ is a locally finite tree with a reference point ϕ. In particular, $(T^{(2)}, \mathcal{A}^{(2)}, \phi)$ is called the infinite binary tree.

It is easy to see that for any infinite geodesic ray $(w_0, w_1, \ldots)$, there exists a geodesic ray originated from ϕ that is equivalent to $(w_0, w_1, \ldots)$. In fact, adding the geodesic $\overline{\phi w_0}$ to $(w_0, w_1, \ldots)$ and removing a loop, one can obtain the infinite geodesic ray having required property. This fact shows the following proposition.

Proposition 2.1.4 *There exists a natural bijective map from Σ to Σ^ϕ.*

Through this map, we always identify the collection of ends Σ and the collection of infinite geodesic rays originated from ϕ, Σ^ϕ.

Hereafter in this monograph, $(T, \mathcal{A}, \phi)$ is assumed to be a locally finite tree with a fixed reference point $\phi \in T$. If no confusion can occur, we omit ϕ in the notations. For example, we use Σ, and Σ_v in place of Σ^ϕ and Σ_v^ϕ respectively.

Example 2.1.5 Let $N \geq 2$ be an integer. In the case of $(T^{(N)}, \mathcal{A}^{(N)}, \phi)$ defined in Example 2.1.3, the collection of the ends Σ is $\Sigma^{(N)} = \{1, \ldots, N\}^{\mathbb{N}} = \{i_1 i_2 i_3 \ldots, | i_j \in \{1, \ldots, N\}$ for any $m \in \mathbb{N}\}$. With the natural product topology, $\Sigma^{(N)}$ is a Cantor set, i.e. perfect and totally disconnected.

Definition 2.1.6 Let $(T, \mathcal{A}, \phi)$ be a locally finite tree with a reference point ϕ.

(1) For $\omega = (w_0, w_1, \ldots) \in \Sigma$, we define $[\omega]_m$ by $[\omega]_m = w_m$ for any $m \geq 0$. Moreover, let $w \in T$. If $\overline{\phi w} = (w_0, w_1, \ldots, w_{|w|})$, then for any $0 \leq m \leq |w|$, we define $[w]_m = w_m$. For $w \in T$, we define

$$T_w = \{v | v \in T, w \in \overline{\phi v}\}.$$

(2) For $w, v \in T$, we define the confluence of w and v, $w \wedge v$, by

$$w \wedge v = w_{\max\{i | i = 0, \ldots, |w|, [w]_i = [v]_i\}}$$

(3) For $\omega, \tau \in \Sigma$, if $\omega \neq \tau$, we define the confluence of ω and τ, $\omega \wedge \tau$, by

$$\omega \wedge \tau = [\omega]_{\max\{m | [\omega]_m = [\tau]_m\}}.$$

(4) For $\omega, \tau \in \Sigma$, we define $\rho_*(\omega, \tau) \geq 0$ by

$$\rho_*(\omega, \tau) = \begin{cases} 2^{-|\omega \wedge \tau|} & \text{if } \omega \neq \tau, \\ 0 & \text{if } \omega = \tau. \end{cases}$$

It is easy to see that ρ_* is a metric on Σ and $\{\Sigma_{[\omega]_m}\}_{m \geq 0}$ is a fundamental system of neighborhood of $\omega \in \Sigma$. Moreover, $\{\Sigma_v\}_{v \in T}$ is a countable base of open sets. This base of open sets has the following property.

Lemma 2.1.7 *Let $(T, \mathcal{A}, \phi)$ be a locally finite tree with a reference point ϕ. Let $w, v \in T$. Then the following three conditions are equivalent to each other.*

(1) $\Sigma_w \cap \Sigma_v \neq \emptyset$
(2) $|w \wedge v| = |w|$ *or* $|w \wedge v| = |v|$.
(3) $\Sigma_v \subseteq \Sigma_w$ *or* $\Sigma_w \subseteq \Sigma_v$

Proof (2) $\Rightarrow$ (3): If $|w \wedge v| = |w|$, then $w = w \wedge v$ and hence $w \in \overline{\phi v}$. Therefore $\Sigma_v \subseteq \Sigma_w$. Similarly if $|w \wedge v| = |v|$, then $\Sigma_w \subseteq \Sigma_v$.

(3) $\Rightarrow$ (1): This is obvious.

(1) $\Rightarrow$ (2): If $\omega \in \Sigma_w \cap \Sigma_v$, then there exist $m, n \geq 0$ such that $w = [\omega]_m$ and $v = [\omega]_m$. It follows that

$$w \wedge v = \begin{cases} w & \text{if } m \leq n, \\ v & \text{if } m \leq n. \end{cases}$$

Hence we see that $|w \wedge v| = |w|$ or $|w \wedge v| = |v|$. $\square$

With the help of the above lemma, we may easily verify the following well-known fact. The proof is standard and left to the readers.

Proposition 2.1.8 *If $(T, \mathcal{A}, \phi)$ is a locally finite tree with a reference point ϕ. Then $\rho_*(\cdot, \cdot)$ is a metric on Σ and (Σ, ρ) is compact and totally disconnected. Moreover, if $\#(S(w)) \geq 2$ for any $w \in T$, then (Σ, ρ) is perfect.*

By the above proposition, if $\#(S(w)) \geq 2$ for any $w \in T$, then Σ is (homeomorphic to) the Cantor set.

2.2   Partition

In this section, we formulate exactly the notion of a partition introduced in Sect. 1.1. A partition is a map from a tree to the collection of nonempty compact subsets of a compact metrizable topological space with no isolated point and it is required to preserve the natural hierarchical structure of the tree. Consequently, a partition induces a surjective map from the Cantor set, i.e. the collection of ends of the tree, to the compact metrizable space.

Throughout this section, $\mathcal{T} = (T, A, \phi)$ is a locally finite tree with a reference point ϕ.

Definition 2.2.1 (Partition) Let $(X, \mathcal{O})$ be a compact metrizable topological space having no isolated point, where $\mathcal{O}$ is the collection of open sets, and let $\mathcal{C}(X, \mathcal{O})$ be the collection of nonempty compact subsets of X. If no confusion can occur, we write $\mathcal{C}(X)$ in place of $\mathcal{C}(X, \mathcal{O})$.

(1) A map $K : T \to \mathcal{C}(X, \mathcal{O})$, where we customarily denote $K(w)$ by K_w for simplicity, is called a partition of X parametrized by $(T, \mathcal{A}, \phi)$ if and only if it satisfies the following conditions (P1) and (P2), which correspond to (1.1.3) and (1.1.4) respectively.

(P1) $K_\phi = X$ and for any $w \in T$, $\#(K_w) \geq 2$, K_w has no isolated point and

$$K_w = \bigcup_{v \in S(w)} K_v.$$

(P2) For any $\omega \in \Sigma$, $\cap_{m \geq 0} K_{[\omega]_m}$ is a single point.

(2) Let $K : T \to \mathcal{C}(X, \mathcal{O})$ be a partition of X parametrized by $(T, \mathcal{A}, \phi)$. Define O_w and B_w for $w \in T$ by

$$O_w = K_w \backslash \left(\bigcup_{v \in (T)_{|w|} \backslash \{w\}} K_v \right),$$

$$B_w = K_w \cap \left(\bigcup_{v \in (T)_{|w|} \backslash \{w\}} K_v \right).$$

If $O_w \neq \emptyset$ for any $w \in T$, then the partition K is called minimal.

(3) Let $K : T \to \mathcal{C}(X, \mathcal{O})$ be a partition of X. Then $(w(1), \ldots, w(m)) \in \cup_{k \geq 0} T^k$ is called a chain of K (or a chain for short if no confusion can occur) if and only if $K_{w(i)} \cap K_{w(i+1)} \neq \emptyset$ for any $i = 1, \ldots, m-1$. A chain $(w(1), \ldots, w(m))$ of K is called a chain of K in $\Lambda \subseteq T$ if $w(i) \in \Lambda$ for any $i = 1, \ldots, m$. For subsets $A, B \subseteq X$, A chain $(w(1), \ldots, w(m))$ of K is called a chain of K between A and B if and only if $A \cap K_{w(1)} \neq \emptyset$ and $B \cap K_{w(m)} \neq \emptyset$. We use $\mathcal{CH}_K(A, B)$ to denote the collection of chains of K between A and B. Moreover, we denote the collection of chains of K in Λ between A and B by $\mathcal{CH}_K^\Lambda(A, B)$.

As is shown in Theorem 2.2.9, a partition can be modified so as to be minimal by restricting it to a suitable subtree.

The next lemma is an assortment of direct consequences from the definition of the partition.

Lemma 2.2.2 *Let $K : T \to \mathcal{C}(X, \mathcal{O})$ be a partition of X parametrized by (T, A, ϕ).*

(1) *For any $w \in T$, O_w is an open set. $O_v \subseteq O_w$ for any $v \in S(w)$.*
(2) *$O_w \cap K_v = \emptyset$ if $w, v \in T$ and $\Sigma_w \cap \Sigma_v = \emptyset$.*
(3) *If $\Sigma_w \cap \Sigma_v = \emptyset$, then $K_w \cap K_v = B_w \cap B_v$.*

Proof (1) Note that by (P1), $X = \cup_{w \in (T)_m} K_w$. Hence

$$O_w = K_w \backslash (\cup_{v \in (T)_{|w|} \backslash \{w\}} K_v) = X \backslash (\cup_{v \in (T)_{|w|} \backslash \{w\}} K_v).$$

So O_w is open. The rest of the statement is immediate from the property (P2).

(2) By Lemma 2.1.7, if $u = w \wedge v$, then $|u| < |w|$ and $|u| < |v|$. Let $w' = [w]_{|u|+1}$ and let $v' = [v]_{|u|+1}$. Then $w', v' \in S(u)$ and $w' \neq v'$. Since $O_{w'} \subseteq K_{w'} \backslash K_{v'}$, it follows that $O_{w'} \cap K_{v'} = \emptyset$. Using (1), we see $O_w \cap K_v = \emptyset$.

(3) This follows immediately by the definition of B_w. $\square$

The condition (P2) provides a natural map from the ends of the tree Σ to the space X.

Proposition 2.2.3 *Let* $K : T \to \mathcal{C}(X, \mathcal{O})$ *be a partition of* X *parametrized by* $(T, \mathcal{A}, \phi)$.

(1) *For* $\omega \in \Sigma$, *define* $\sigma(\omega)$ *as the single point* $\cap_{m \geq 0} K_{[\omega]_m}$. *Then* $\sigma : \Sigma \to X$ *is continuous and surjective. Moreover,* $\sigma(\Sigma_w) = K_w$ *for any* $w \in T$.

(2) *The partition* $K : T \to \mathcal{C}(X, \mathcal{O})$ *is minimal if and only if* K_w *is the closure of* O_w *for any* $w \in T$. *Moreover, if* $K : T \to \mathcal{C}(X, \mathcal{O})$ *is minimal then* O_w *coincides with the interior of* K_w.

Proof (1) Note that $K_w = \cup_{v \in S(w)} K_v$. Hence if $x \in K_w$, then there exists $v \in S(w)$ such that $x \in K_v$. Using this fact inductively, we see that, for any $x \in X$, there exists $\omega \in \Sigma$ such that $x \in K_{[\omega]_m}$ for any $m \geq 0$. Since $x \in \cap_{m \geq 0} K_{[\omega]_m}$, (P2) shows that $\sigma(\omega) = x$. Hence σ is surjective. At the same time, it follows that $\sigma(\Sigma_w) = K_w$. Let U be an open set in X. For any $\omega \in \sigma^{-1}(U)$, $K_{[\omega]_m} \subseteq U$ for sufficiently large m. Then $\Sigma_{[\omega]_m} \subseteq \sigma^{-1}(U)$. This shows that $\sigma^{-1}(U)$ is an open set and hence σ is continuous.

(2) Let $\overline{O}_w$ be the closure of O_w. If $K_w = \overline{O}_w$ for any $w \in T$, then $O_w \neq \emptyset$ for any $w \in T$ and hence $K : T \to \mathcal{C}(X, \mathcal{O})$ is minimal. Conversely, assume that $K : T \to \mathcal{C}(X, \mathcal{O})$ is minimal. By Lemma 2.2.2, $\overline{O}_{[\omega]_m} \supseteq \overline{O}_{[\omega]_{m+1}}$ for any $\omega \in \Sigma$ and $m \geq 0$. Hence $\{\sigma(\omega)\} = \cap_{m \geq 0} K_{[\omega]_m} = \cap_{m \geq 0} \overline{O}_{[\omega]_m} \subseteq \overline{O}_{[\omega]_n}$ for any $n \geq 0$. This yields that $\sigma(\Sigma_w) \subseteq \overline{O}_w$. Since $\sigma(\Sigma_w) = K_w$, this implies $\overline{O}_w = K_w$. Now if K is minimal, since O_w is open by Lemma 2.2.2-(1), it follows that O_w is the interior of K_w. $\square$

Definition 2.2.4 A partition $K : T \to \mathcal{C}(X, \mathcal{O})$ parametrized by a tree $(T, \mathcal{A}, \phi)$ is called strongly finite if and only if

$$\sup_{x \in X} \#(\sigma^{-1}(x)) < +\infty,$$

where $\sigma : \Sigma \to X$ is the map defined in Proposition 2.2.3-(1).

Definition 2.2.5 Let (X, d) be a metric space. (X, d) is said to be (geometrically) doubling if and only if there exists $N > 0$ such that for any $x \in X$ and $r > 0$, there exist $x_1, \dots, x_N \in X$ such that $B_d(x, r) \subseteq \cup_{i=1,\dots,N} B_d(x_i, r/2)$.

For geometrically doubling metric spaces, one can always construct partitions. In fact, systems of dyadic cubes constructed by Hytönen and Kairema in [17] are minimal partitions. More precisely, their systems of closed dyadic cubes $\{\bar{Q}_\alpha^k\}$ and open dyadic cubes $\{\tilde{Q}_\alpha^k\}$ correspond to a partition $\{K_w\}$ and the collection of its interiors $\{O_w\}$ respectively. Detailed examination of [17] leads to the following theorem.

Theorem 2.2.6 (Hytönen and Kairema) *If* (X, d) *be a compact metric space which has no isolated point. Assume that* (X, d) *is geometrically doubling. Then* *there exist $r \in (0, 1)$, $c_1, c_2 > 0$, a tree (T, A, ϕ) and a strongly finite minimal partition K of X parametrized by T such that*

$$\max_{w \in T} \#(S(w)) < \infty$$

and, for any $w \in T$,

$$B_d(x_w, c_1 r^{|w|}) \subseteq K_w \subseteq B_d(x_w, c_2 r^{|w|})$$

for some $x_w \in K_w$.

Example 2.2.7 Let (Y, d) be a complete metric space and let $\{F_1, \ldots, F_N\}$ be collection of contractions from (Y, d) to itself, i.e. $F_i : Y \to Y$ and

$$\sup_{x \neq y \in Y} \frac{d(F_i(x), F_i(y))}{d(x, y)} < 1$$

for any $i = 1, \ldots, N$. Then it is well-known that there exists a unique nonempty compact set X such that

$$X = \bigcup_{i=1,\ldots,N} F_i(X).$$

See [19, Section 1.1] for a proof of this fact for example. X is called the self-similar set associated with $\{F_1, \ldots, F_N\}$. Let $(T^{(N)}, \mathcal{A}^{(N)}, \phi)$ be the tree defined in Example 2.1.3. For any $i_1 \ldots i_m \in T$, set $F_{i_1 \ldots i_m} = F_{i_1} \circ \ldots \circ F_{i_m}$ and define $K_w = F_w(X)$. Then $K : T^{(N)} \to C(X)$ is a partition of K parametrized by $(T^{(N)}, \mathcal{A}^{(N)}, \phi)$. See [19, Section 1.2]. The associated map from $\Sigma = \{1, \ldots, N\}^{\mathbb{N}}$ to K is sometimes called the coding map. To determine if K is minimal or not is known to be rather delicate issue. See [19, Theorem 1.3.8] for example.

Example 2.2.8 (The Sierpinski Carpet, Fig. 2.1) This is the special case of self-similar sets presented in the last example. Let $p_1 = (0, 0)$, $p_2 = (\frac{1}{2}, 0)$, $p_3 = (1, 0)$, $p_4 = (1, \frac{1}{2})$, $p_5 = (1, 1)$, $p_6 = (\frac{1}{2}, 1)$, $p_7 = (0, 1)$ and $p_8 = (0, \frac{1}{2})$. Set $F_i : [0, 1]^2 \to [0, 1]^2$ for $i = 1, \ldots, 8$ by

$$F_i(x) = \frac{1}{3}(x - p_i) + p_i$$

for any $x \in [0, 1]^2$. The unique nonempty compact set X satisfying

$$X = \bigcup_{i=1}^{8} F_i(X)$$

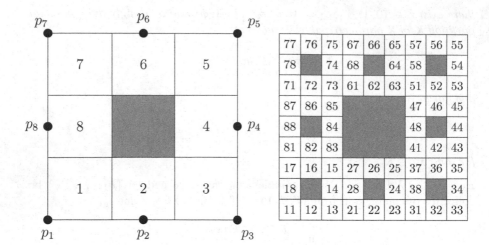

Fig. 2.1 Partition: the Sierpinski carpet

is called the Sierpinski carpet. In this case, the associated tree is $(T^{(8)}, \mathcal{A}^{(8)}, \phi)$. Define $K : T^{(8)} \to \mathcal{C}(X, \mathcal{O})$ by

$$K_{i_1 \ldots i_m} = F_{i_1 \ldots i_m}(X)$$

as in Example 2.2.7. Then K is a partition of X parametrized by the tree $(T^{(8)}, \mathcal{A}^{(8)}, \phi)$. In Fig. 2.1, K_{ij} is represented by ij for simplicity.

Removing unnecessary vertices of the tree, we can always modify the original partition and obtain a minimal one.

Theorem 2.2.9 *Let* $K : T \to \mathcal{C}(X, \mathcal{O})$ *be a partition of* X *parametrized by* $(T, \mathcal{A}, \phi)$. *There exist* $T' \subseteq T$ *and* $K' : T' \to \mathcal{C}(X, \mathcal{O})$ *such that* $(T', \mathcal{A}|_{T' \times T'})$ *is a tree,* $\phi \in T'$, $K'_w \subseteq K_w$ *for any* $w \in T'$ *and* K' *is a minimal partition of* X *parametrized by* $(T', \mathcal{A}', \phi)$.

To prove the above theorem, we need the following lemma.

Lemma 2.2.10 *Let* X *be a set. A collection of the subsets* $\{X_1, \ldots, X_m\}$ *of* X *is called a minimal covering of* X *if and only if* $X = \cup_{i=1}^m X_i$ *and* $X_i \setminus \cup_{j : j \neq i} X_j \neq \emptyset$ *for any* $i = 1, \ldots, m$. *Let* $\{K_1, \ldots, K_n\}$ *be a collection of subsets of* X. *If* $K_i \neq \emptyset$ *for any* $i = 1, \ldots, n$ *and* $X = \cup_{i=1}^n K_i$, *then there exists a minimal covering* $\{K_{i_1}, \ldots, K_{i_k}\} \subseteq \{K_1, \ldots, K_m\}$.

Proof If $\{K_1, \ldots, K_n\}$ is not a minimal covering, then there exists i such that $K_i \subseteq \cup_{j : j \neq i} K_j$. Removing K_i form $\{K_1, \ldots, K_n\}$, we obtain a covering of X which contains $n - 1$ elements. Repeating this procedure, we will eventually obtain a minimal covering. $\qquad\square$

Proof of Theorem 2.2.9 We are going to conduct inductive construction of a sequence $\{T^{(m)}\}_{m\geq 0}$ of subtrees of T and $\{K_w^{(m)}\}_{w\in T^{(m)}}$ possessing the following properties (a), (b), (c), (d) and (e):

(a) $\{K_w^{(m)}\}_{w\in T^{(m)}}$ is a partition of X parametrized by $T^{(m)}$.
(b) $\{K_w^{(m)}\}_{w\in(T^{(m)})_i}$ is a minimal covering of X for any $i = 0,\ldots,m$.
(c) $T^{(m+1)} \subseteq T^{(m)}$ and $K_w^{(m+1)} \subseteq K_w^{(m)}$ for any $w \in T^{(m+1)}$
(d) $(T^{(m+1)})_i = (T^{(m)})_i$ for any $i = 0,\ldots,m$
(e) $O_w^{(m+1)} \supseteq O_w^{(m)}$ for any $w \in \cup_{i=0}^m (T^{(m)})_i$.

First let $T^{(0)} = T$ and $K_w^{(0)} = K_w$ for any $w \in T$. Suppose that we have constructed a sequence $T^{(0)}, T^{(1)}, \ldots, T^{(m)}$ with the properties (a), (b), (c), (d) and (e). Then by Lemma 2.2.10, there exists a subset A of $(T^{(m)})_{m+1}$ such that $\{K_w^{(m)}\}_{w\in A}$ is a minimal covering of X. Set $(T^{(m+1)})_{m+1} = A$. Define

$$(T^{(m+1)})_k = \begin{cases} (T^{(m)})_k & \text{for } k = 1,\ldots,m \\ (T^{(m)})_k \cap (\cup_{w\in(T^{(m+1)})_{m+1}} T_w^{(m)}) & \text{for } k \geq m+1 \end{cases}$$

and

$$K_w^{(m+1)} = \begin{cases} \bigcup_{v\in(T^{(m+1)})_{m+1}\cap T_w^{(m)}} K_v^{(m)} & \text{if } |w| \leq m \\ K_w^{(m)} & \text{if } |w| \geq m+1 \end{cases}.$$

Then (c) and (d) are obvious by definition. To verify (a), if $i \leq m$, we have

$$\bigcup_{w\in(T^{(m+1)})_i} K_w^{(m+1)} = \bigcup_{w\in(T^{(m)})_i} \left(\bigcup_{v\in(T^{(m+1)})_{m+1}\cap T_w^{(m)}} K_v \right)$$

$$= \bigcup_{v\in(T^{(m+1)})_{m+1}} K_v^{(m+1)} = X.$$

If $i \geq m+1$, since $\{K_w^{(m+1)}\}_{w\in(T^{(m+1)})_{m+1}}$ is a minimal covering,

$$\bigcup_{w\in(T^{(m+1)})_i} K_w^{(m+1)} = \bigcup_{v\in(T^{(m+1)})_{m+1}} \left(\bigcup_{w\in T_v^{(m+1)}\cap(T^{(m+1)})_i} K_w^{(m+1)} \right)$$

$$\bigcup_{v\in(T^{(m+1)})_{m+1}} K_v^{(m+1)} = X.$$

The condition (P2) is straightforward from (c). Also by (c), if $w \in T^{(m)}$ and $|w| \leq m$, then

$$O_w^{(m)} = X \backslash \left(\bigcup_{v \in (T^{(m)})_{|w|}, v \neq w} K_v^{(m)} \right) \subseteq X \backslash \left(\bigcup_{v \in (T^{(m)})_{|w|}, v \neq w} K_v^{(m+1)} \right) = O_w^{(m+1)}.$$

This shows (d) and (b). Consequently, we have constructed $\{T^{(k)}\}_{k=0,1,\dots,m+1}$ with the properties (a) through (e). Thus we have obtained inductively a sequence $\{T^{(m)}\}_{m \geq 0}$ with the required properties. Now define T' and $\{K'_w\}_{w \in T'}$ by $T' = \cap_{m \geq 0} T^{(m)}$ and $K'_w = \cap_{m \geq 0} K_w^{(m)}$. Then by (a), (b), (c), (d) and (e), we see that K' is a minimal partition parametrized by T'.   □

A partition $K : T \to \mathcal{C}(X, \mathcal{O})$ induces natural graph structure on T. In the rest of this section, we show that T can be regarded as the hyperbolic filling of X if the induced graph structure is hyperbolic. See [9], for example, about the notion of hyperbolic fillings.

Definition 2.2.11 Let $K : T \to \mathcal{C}(X, \mathcal{O})$ be a partition. Then define

$$E_m^h = \{(w, v) | w, v \in (T)_m, w \neq v, K_w \cap K_v \neq \emptyset\}$$

and

$$E^h = \bigcup_{m \geq 0} E_m^h.$$

An element $(u, v) \in E^h$ is called a horizontal edge associated with $(T, \mathcal{A}, \phi)$ and $K : T \to \mathcal{C}(X, \mathcal{O})$. The symbol "h" in the notation E_m^h and E^h represents the word "horizontal". On the contrary, an element $(w, v) \in \mathcal{A}$ is called a vertical edge. Moreover we define

$$\mathcal{B}(w, v) = \begin{cases} 1 & \text{if } \mathcal{A}(w, v) = 1 \text{ or } (w, v) \in E^h, \\ 0 & \text{otherwise.} \end{cases}$$

The graph $(T, \mathcal{B})$ is called the resolution of X associated with the partition $K : T \to \mathcal{C}(X, \mathcal{O})$. We use $d_{(T, \mathcal{B})}(\cdot, \cdot)$ to denote the shortest path metric, i.e.

$$d_{(T, \mathcal{B})}(w, v) = \min\{n | \text{there exists } (w(1), \dots, w(n+1)) \in T^{n+1}$$

$$w = w(1), v = w(n+1), \mathcal{B}(w(i), w(i+1)) = 1 \text{ for any } i = 1, \dots, n\}.$$

Moreover, an infinite sequence $(w(0), w(1), \dots)$ is called a geodesic ray of $(T, \mathcal{B})$ starting from $w(0)$ if and only if $d_{(T, \mathcal{B})}(w(0), w(n)) = n$ for any $n \geq 1$.

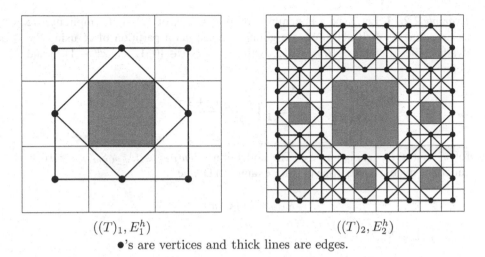

$$((T)_1, E_1^h)$$ $$((T)_2, E_2^h)$$

●'s are vertices and thick lines are edges.

Fig. 2.2 Horizontal edges: the Sierpinski carpet

Remark The horizontal graph $((T)_m, E_m^h)$ is not necessarily connected. More precisely, $((T)_m, E_m^h)$ is connected for any $m \geq 0$ if and only if X is connected. Note that if X is homeomorphic to the Cantor set, then $E_m^h = \emptyset$ for any $m \geq 0$.

In Fig. 2.2, we present $((T)_1, E_1^h)$ and $((T)_2, E_2^h)$ for the Sierpinski carpet introduced in Example 2.2.8.

We will show in Lemma 2.5.1 that if $(\phi, w(1), w(2), \ldots)$ is an infinite geodesic ray in $(T, \mathcal{B})$ with respect to the metric $d_{(T,\mathcal{B})}$ starting from ϕ, then it coincides with $(\phi, [\omega]_1, [\omega]_2, \ldots)$ for some $\omega \in \Sigma$. In other words, the collection of geodesic rays of $(T, \mathcal{B})$ starting from ϕ can be identified with Σ. The following proposition will be proven in Sect. 2.5.

Proposition 2.2.12 *Let* $\omega, \tau \in \Sigma$. $\sup_{n \geq 1} d_{(T,\mathcal{B})}([\omega]_n, [\tau]_n) < +\infty$ *if and only if* $\sigma(\omega) = \sigma(\tau)$.

By this proposition, whether the resolution $(T, \mathcal{B})$ is hyperbolic or not, X can be identified with the quotient space of the geodesic rays under the equivalence relation $\sim$ defined as $\omega \sim \tau$ if and only if $\sup_{n \geq 1} d_{(T,\mathcal{B})}([\omega]_n, [\tau]_n) < +\infty$. In case of a hyperbolic graph, such a quotient space has been called the hyperbolic boundary of the graph in the framework of Gromov theory of hyperbolic metric spaces. We will give detailed accounts on these points later in Sect. 2.5.

In [15], Elek has constructed a hyperbolic graph whose hyperbolic boundary is homeomorphic to a given compact subset of $\mathbb{R}^N$. From our point of view, what he has done is to construct a partition of the compact metric space using dyadic cubes as is seen in the next example. However, the resolution $(T, \mathcal{B})$ associated with the partition is slightly different from the original graph constructed by Elek. See the details below.

Example 2.2.13 Let X be a nonempty compact subset of $\mathbb{R}^N$. For simplicity, we assume that $X \subseteq [0, 1]^N$. We are going to construct a partition of X using the dyadic cubes. Let $S_m = \{(m, i_1, \ldots, i_N) | (i_1, \ldots, i_N) \in \{0, 1, \ldots, 2^m - 1\}^N\}$ and define

$$C(w) = \prod_{j=1}^{N} \left[\frac{i_j}{2^m}, \frac{i_j + 1}{2^m}\right]$$

for $w = (m, i_1, \ldots, i_N) \in S_m$. The collection $\{C(w) | w \in \cup_{m \geq 0} S_m, m \geq 0\}$ is called the dyadic cubes. (See [14] for example.) Define

$$K_w = X \cap C(w)$$

for $w \in \cup_{m \geq 0} S_m$,

$$(T)_m = \{w | w \in S_m, K_w \neq \emptyset\}$$

for $m \geq 0$ and $T = \cup_{m \geq 0}(T)_m$. Moreover, we define $(w, v) \in \mathcal{A}$ for $(w, v) \in T \times T$ if $(|w| = |v| + 1$ and $C(w) \subseteq C(v))$ or $(|v| = |w| + 1$ and $C(v) \subseteq C(w))$. Then $(T, \mathcal{A}, \phi)$ is a tree with a reference point ϕ, where $\phi = (0, 0, \ldots, 0) \in (T)_0$ and the map from $w \in T$ to $K_w \in \mathcal{C}(X, \mathcal{O})$ is a partition of X parametrized by $(T, \mathcal{A}, \phi)$. The hyperbolic graph constructed by Elek in [15] is a slight modification of the resolution $(T, \mathcal{B})$. In fact, the vertical edges are the same but Elek's graph has more horizontal edges. Precisely set

$$\widetilde{E}_m^h = \{(w, v) | w, v \in (T)_m, w \neq v, C(w) \cap C(v) \neq \emptyset\}.$$

and define $\widetilde{\mathcal{B}} = \mathcal{A} \cup (\cup_{m \geq 1} \widetilde{E}_m^h)$. Then Elek's graph coincides with $(T, \widetilde{\mathcal{B}})$. Note that in $(T, \mathcal{B})$, the horizontal edges are

$$E_m^h = \{(w, v) | w, v \in (T)_m, w \neq v, K_w \cap K_v \neq \emptyset\}.$$

So, $\widetilde{E}_m^h \supseteq E_m^h$ and hence $\widetilde{\mathcal{B}} \supseteq \mathcal{B}$ in general. In Example 2.5.18, we are going to show that $(T, \mathcal{B})$ is hyperbolic as a corollary of our general framework.

2.3   Weight Function and Associated "Visual Pre-metric"

Throughout this section, $(T, \mathcal{A}, \phi)$ is a locally finite tree with a reference point ϕ, $(X, \mathcal{O})$ is a compact metrizable topological space with no isolated point and $K : T \to \mathcal{C}(X, \mathcal{O})$ is a partition of X parametrized by $(T, \mathcal{A}, \phi)$.

In this section, we introduce the notion of a weight function, which assigns each vertex of the tree T a "size" or "weight". Then, from the weight function, we construct a kind of "balls" and a "distance" on the compact metric space X.

Definition 2.3.1 (Weight Function) A function $g : T \to (0, 1]$ is called a weight function if and only if it satisfies the following conditions (G1), (G2) and (G3):

(G1) $g(\phi) = 1$
(G2) For any $w \in T$, $g(\pi(w)) \geq g(w)$
(G3) $\lim_{m \to \infty} \sup_{w \in (T)_m} g(w) = 0$.

We denote the collection of all the weight functions by $\mathcal{G}(T)$. Let g be a weight function. We define

$$\Lambda_s^g = \{w | w \in T, g(\pi(w)) > s \geq g(w)\}$$

for any $s \in (0, 1)$. $\{\Lambda_s^g\}_{s \in (0,1]}$ is called the scale associated with g. For $s \geq 1$, we define $\Lambda_s^g = \{\phi\}$.

Remark To be exact, one should use $\mathcal{G}(T, A, \phi)$ rather than $\mathcal{G}(T)$ as the notation for the collection of all the weight functions because the notion of weight function apparently depends not only on the set T but also the structure of T as a tree. We use, however, $\mathcal{G}(T)$ for simplicity as long as no confusion may occur.

Remark In the case of the partitions associated with a self-similar set appearing in Example 2.2.7, the counterpart of weight functions was called gauge functions in [20]. Also $\{\Lambda_s^g\}_{0 < s \leq 1}$ was called the scale associated with the gauge function g.

Given a weight function g, we consider $g(w)$ as a virtual "size" or "diameter" of Σ_w for each $w \in T$. The set Λ_s^g is the collection of subsets Σ_w's whose sizes are approximately s.

Proposition 2.3.2 *Suppose that* $g : T \to (0, 1]$ *satisfies* (G1) *and* (G2). g *is a weight function if and only if*

$$\lim_{m \to \infty} g([\omega]_m) = 0 \tag{2.3.1}$$

for any $\omega \in \Sigma$.

Proof If g is a weight function, i.e. (G3) holds, then (2.3.1) is immediate.
   Suppose that (G3) does not hold, i.e. there exists $\epsilon > 0$ such that

$$\sup_{w \in (T)_m} g(w) > \epsilon \tag{2.3.2}$$

for any $m \geq 0$. Define $Z = \{w | w \in T, g(w) > \epsilon\}$ and $Z_m = (T)_m \cap Z$. By (2.3.2), $Z_m \neq \emptyset$ for any $m \geq 0$. Since $\pi(w) \in Z$ for any $w \in Z$, if $Z_{m,n} = \pi^{n-m}(Z_n)$ for any $n \geq m$, where π^k is the k-th iteration of π, then $Z_{m,n} \neq \emptyset$ and $Z_{m,n} \supseteq Z_{m,n+1}$ for any $n \geq m$. Set $Z_m^* = \cap_{n \geq m} Z_{m,n}$. Since $(T)_m$ is a finite set and so is $Z_{m,n}$, we see that $Z_m^* \neq \emptyset$ and $\pi(Z_{m+1}^*) = Z_m^*$ for any $m \geq 0$. Note that $Z_0^* = \{\phi\}$. Inductively, we may construct a sequence $(\phi, w(1), w(2), \ldots)$ satisfying $\pi(w(m+1)) = w(m)$ and $w(m) \in Z_m^*$ for any $m \geq 0$. Set $\omega = (\phi, w(1), w(2), \ldots)$. Then $\omega \in \Sigma$ and $g([\omega]_m) \geq \epsilon$ for any $m \geq 0$. This contradicts (2.3.1).   □

Proposition 2.3.3 *Let $g : T \to (0, 1]$ be a weight function and let $s \in (0, 1]$. Then*

$$\bigcup_{w \in \Lambda_s^g} \Sigma_w = \Sigma \tag{2.3.3}$$

and if $w, v \in \Lambda_s^g$ and $w \neq v$, then

$$\Sigma_w \cap \Sigma_v = \emptyset.$$

Proof For any $\omega = (w_0, w_1, \ldots) \in \Sigma$, $\{g(w_i)\}_{i=0,1,\ldots}$ is monotonically non-increasing sequence converging to 0 as $i \to \infty$. Hence there exists a unique $m \geq 0$ such that $g(w_{m-1}) > s \geq g(w_m)$. Therefore, there exists a unique $m \geq 0$ such that $[\omega]_m \in \Lambda_s^g$. Now (2.3.3) is immediate. Assume $w, v \in \Lambda_s^g$ and $\Sigma_v \cap \Sigma_w \neq \emptyset$. Choose $\omega = (w_0, w_1, \ldots) \in \Sigma_v \cap \Sigma_w$. Then there exist $m, n \geq 0$ such that $[\omega]_m = w_m = w$ and $[\omega]_n = w_n = v$. By the above fact, we have $m = n$ and hence $w = v$.   □

By means of the partition $K : T \to \mathcal{C}(X, \mathcal{O})$, one can define weight functions naturally associated with metrics and measures on X as follows.

Notation Let d be a metric on X. We define the diameter of a subset $A \subseteq X$ with respect to d, $\mathrm{diam}(A, d)$ by $\mathrm{diam}(A, d) = \sup\{d(x, y) | x, y \in A\}$. Moreover, for $x \in X$ and $r > 0$, we set $B_d(x, r) = \{y | y \in X, d(x, y) < r\}$.

Definition 2.3.4

(1) Define

$$\mathcal{D}(X, \mathcal{O}) = \{d | d \text{ is a metric on } X \text{ inducing the topology } \mathcal{O} \text{ and}$$

$$\mathrm{diam}(X, d) = 1\}.$$

For $d \in \mathcal{D}(X, \mathcal{O})$, define $g_d : T \to (0, 1]$ by $g_d(w) = \mathrm{diam}(K_w, d)$ for any $w \in T$.

(2) Define

$$\mathcal{M}_P(X, \mathcal{O}) = \{\mu | \mu \text{ is a Radon probability measure on } (X, \mathcal{O})$$

$$\text{satisfying } \mu(\{x\}) = 0 \text{ for any } x \in X \text{ and } \mu(K_w) > 0 \text{ for any } w \in T\}.$$

For $\mu \in \mathcal{M}_P(X, \mathcal{O})$, define $g_\mu : T \to (0, 1]$ by $g_\mu(w) = \mu(K_w)$ for any $w \in T$.

The condition $\text{diam}(X, d) = 1$ in the definition of $\mathcal{D}(X, \mathcal{O})$ is only for the purpose of normalization. Note that since $(X, \mathcal{O})$ is compact, if a metric d on X induces the topology $\mathcal{O}$, then $\text{diam}(X, d) < +\infty$.

Proposition 2.3.5

(1) *For any* $d \in \mathcal{D}(X, \mathcal{O})$, g_d *is a weight function.*
(2) *For any* $\mu \in \mathcal{M}_P(X, \mathcal{O})$, g_μ *is a weight function.*

Proof (1) The properties (G1) and (G2) are immediate from the definition of g_d. Suppose that there exists $\omega \in \Sigma$ such that

$$\lim_{m \to \infty} g_d([\omega]_m) > 0. \tag{2.3.4}$$

Let ϵ be the above limit. Since $g_d([\omega]_m) = \text{diam}(K_{[\omega]_m}, d) > \epsilon$, there exist $x_m, y_m \in K_{[\omega]_m}$ such that $d(x_m, y_m) \geq \epsilon$. Note that $K_{[\omega]_m} \supseteq K_{[\omega]_{m+1}}$ and hence $x_n, y_n \in K_{[\omega]_m}$ if $n \geq m$. Since X is compact, there exist subsequences $\{x_{n_i}\}_{i \geq 1}$, $\{y_{n_i}\}_{i \geq 1}$ converging to x and y as $i \to \infty$ respectively. It follows that $x, y \in \cap_{m \geq 0} K_{[\omega]_m}$ and $d(x, y) \geq \epsilon > 0$. This contradicts (P2). Thus we have shown (2.3.1). By Proposition 2.3.2, g_d is a weight function.

(2) As in the case of metrics, (G1) and (G2) are immediate. Let $\omega \in \Sigma$. Then $\cap_{m \geq 0} K_{[\omega]_m} = \{\sigma(\omega)\}$. Therefore, $g_\mu([\omega]_m) = \mu(K_{[\omega]_m}) \to 0$ as $m \to \infty$. Hence we verify (2.3.1). Thus by Proposition 2.3.2, g_μ is a weight function. $\square$

The weight function g_d and g_μ are called the weight functions associated with d and μ respectively. Although the maps $d \to g_d$ and $\mu \to g_\mu$ may not be injective, we sometimes abuse notations and use d and μ to denote g_d and g_μ respectively.

Through a partition we introduce the notion of "balls" of X associated with a weight function.

Definition 2.3.6 Let $g : T \to (0, 1]$ be a weight function.

(1) For $s \in (0, 1]$, $w \in \Lambda_s^g$, $M \geq 0$ and $x \in X$, we define

$$\Lambda_{s,M}^g(w) = \{v | v \in \Lambda_s^g, \text{there exists a chain } (w(1), \ldots, w(k)) \text{ of } K \text{ in } \Lambda_s^g$$

$$\text{such that } w(1) = w, w(k) = v \text{ and } k \leq M + 1\}$$

and

$$\Lambda_{s,M}^g(x) = \bigcup_{w \in \Lambda_s^g \text{ and } x \in K_w} \Lambda_{s,M}^g(w).$$

For $x \in X$, $s \in (0, 1]$ and $M \geq 0$, define

$$U_M^g(x, s) = \bigcup_{w \in \Lambda_{s,M}^g(x)} K_w.$$

We let $U_M^g(x, s) = X$ if $s \geq 1$.

In Fig. 2.3, we show examples of $U_M^g(x, s)$ for the Sierpinski carpet introduced in Example 2.2.8.

The family $\{U_M^g(x, s)\}_{s>0}$ is a fundamental system of neighborhood of $x \in X$ as is shown in Proposition 2.3.7.

Note that

$$\Lambda_{s,0}^g(w) = \{w\} \quad \text{and} \quad \Lambda_{s,1}^g(w) = \{v | v \in \Lambda_s^g, K_v \cap K_w \neq \emptyset\}$$

for any $w \in \Lambda_s^g$ and

$$\Lambda_{s,0}^g(x) = \{w | w \in \Lambda_s^g, x \in K_w\} \quad \text{and} \quad U_0^g(x, s) = \bigcup_{w : w \in \Lambda_s^g, x \in K_w} K_w$$

for any $x \in X$. Moreover,

$$U_M^g(x, s) = \{y | y \in X, \text{there exists } (w(1), \ldots, w(M+1)) \in \mathcal{CH}_K^{\Lambda_s^g}(x, y).\}$$

Proposition 2.3.7 *Let K be a partition of X parametrized by $(T, \mathcal{A}, \phi)$ and let $g : T \to (0, 1]$ be a weight function. For any $s \in (0, 1]$ and $x \in X$, $U_0^g(x, s)$ is a neighborhood of x. Furthermore, $\{U_M^g(x, s)\}_{s \in (0,1]}$ is a fundamental system of neighborhood of x for any $x \in X$.*

Proof Let d be a metric on X giving the original topology of $(X, \mathcal{O})$. Assume that for any $r > 0$, there exists $y \in B_d(x, r)$ such that $y \notin U_0^g(x, s)$. Then there exists a sequence $\{x_n\}_{n \geq 1} \subseteq X$ such that $x_n \to x$ as $n \to \infty$ and $x_n \notin U_0^g(x, s)$ for any $n \geq 1$. Since Λ_s^g is a finite set, there exists $w \in \Lambda_s$ which includes infinite members of $\{x_n\}_{n \geq 1}$. By the closedness of K_w, it follows that $x \in K_w$ and $x_n \in K_w \subseteq U_0^g(x, s)$. This contradiction shows that $U_0^g(x, s)$ contains $B_d(x, r)$ for some $r > 0$.

Next note $\min_{w \in \Lambda_s^g} |w| \to \infty$ as $s \downarrow 0$. This along with that fact that g_d is a weight function implies that $\max_{w \in \Lambda_s^g} \text{diam}(K_w, d) \to 0$ as $s \downarrow 0$. Set $\rho_s = \max_{w \in \Lambda_s^g} \text{diam}(K_w, d)$. Then $\text{diam}(U_M^g(x, s), d) \leq (M+1)\rho_s \to 0$ as $s \downarrow 0$. This implies that $\cap_{s \in (0,1]} U_M^g(x, s) = \{x\}$. Thus $\{U_M^g(x, s)\}_{s \in (0,1]}$ is a fundamental system of neighborhoods of x. $\square$

We regard $U_M^g(x, s)$ as a virtual "ball" of radius s and center x. In fact, there exists a kind of "pre-metric" $\delta_M^g : X \times X \to [0, \infty)$ such that $\delta_M^g(x, y) > 0$ if and only if $x \neq y$, $\delta_M^g(x, y) = \delta_M^g(y, x)$ and

$$U_M^g(x, s) = \{y | \delta_M^g(x, y) \leq s\}. \tag{2.3.5}$$

As is seen in the next section, however, the pre-metric δ_M^g may not satisfy the triangle inequality in general.

Definition 2.3.8 Let $M \geq 0$. Define $\delta_M^g(x, y)$ for $x, y \in X$ by

$$\delta_M^g(x, y) = \inf\{s \mid s \in (0, 1], y \in U_M^g(x, s)\}.$$

$\delta_M^g(\cdot, \cdot)$ is called the visual pre-metric associated with the weight function g and the parameter M.

Remark For any $g \in \mathcal{G}(T)$, $M \geq 0$ and $x \in X$, it follows that $\Lambda_{1,M}^g(x) = \{\phi\}$ and hence $U_M^g(x, 1) = X$. So, $\delta_M^g(x, y) \leq 1$ for any $x, y \in X$.

Our visual pre-metric δ_M^g can be thought of as a counterpart of the "visual metric" in the sense of Bonk-Meyer in [8] and the "visual pre-metric" in the framework of Gromov hyperbolic metric spaces, whose exposition can be found in [12] and [27]. In fact, if certain rearrangement of the resolution $(X, \mathcal{B})$ associated with the weight function is hyperbolic, then δ_M^g is bi-Lipschitz equivalent to a visual pre-metric in the sense of Gromov. See Theorem 2.5.12 for details.

Proposition 2.3.9 *For any $M \geq 0$ and $x, y \in X$,*

$$\delta_M^g(x, y) = \min\{s \mid s \in (0, 1], y \in U_M^g(x, s)\}. \tag{2.3.6}$$

In particular, (2.3.5) holds for any $M \geq 0$ and $s \in (0, 1]$.

Proof The property (G3) implies that for any $t \in (0, 1]$, there exists $n \geq 0$ such that $\cup_{s \geq t} \Lambda_s^g \subseteq \cup_{m=0}^n (T)_m$. Hence $\{(w(1), \ldots, w(M+1)) \mid w(i) \in \cup_{s \geq t} \Lambda_s^g\}$ is finite. Let $s_* = \delta_M^g(x, y)$. Then there exist a sequence $\{s_m\}_{m \geq 1} \subseteq [s_*, 1]$ and $(w_m(1), \ldots, w_m(M+1)) \in (\Lambda_{s_m}^g)^{M+1}$ such that $\lim_{m \to \infty} s_m = s_*$ and $(w_m(1), \ldots, w_m(M+1))$ is a chain between x and y for any $m \geq 1$. Since $\{(w(1), \ldots, w(M+1)) \mid w(i) \in \cup_{s \geq s_*} \Lambda_s^g\}$ is finite, there exists $(w_*(1), \ldots, w_*(M+1))$ such that $(w_*(1), \ldots, w_*(M+1)) = (w_m(1), \ldots, w_m(M+1))$ for infinitely many m. For such m, we have $g(\pi(w_*(i))) > s_m \geq g(w_*(i))$ for any $i = 1, \ldots, M+1$. This implies that $w_*(i) \in \Lambda_{s_*}^g$ for any $i = 1, \ldots, M+1$ and hence $y \in U_M^g(x, s_*)$. Thus we have shown (2.3.6). □

2.4 Metrics Adapted to Weight Function

Now we start to investigate the first question mentioned in the introduction, which is when a weight function is naturally associated with a metric. In this section, however, we only give an answer to a weaker version of the problem, which is Theorem 2.4.12. The original version of the problem will be dealt with in Sect. 4.1, where a sufficient condition for the existence of a natural metric associated with a given weight function will be obtained.

As in the last section, $(T, \mathcal{A}, \phi)$ is a locally finite tree with a reference point ϕ, $(X, \mathcal{O})$ is a compact metrizable topological space with no isolated point and $K : T \to \mathcal{C}(X, \mathcal{O})$ is a partition throughout this section.

The purpose of the next definition is to clarify when the virtual balls $U_M^g(x, s)$ induced by a weight function g can be thought of as real "balls" derived from a metric.

Definition 2.4.1 Let $M \geq 0$. A metric $d \in \mathcal{D}(X, \mathcal{O})$ is said to be M-adapted to g if and only if there exist $\alpha_1, \alpha_2 > 0$ such that

$$U_M^g(x, \alpha_1 r) \subseteq B_d(x, r) \subseteq U_M^g(x, \alpha_2 r)$$

for any $x \in X$ and $r > 0$. d is said to be adapted to g if and only if d is M-adapted to g for some $M \geq 0$.

Now our question is the existence of a metric adapted to a given weight function. The number M really makes a difference in the above definition. Namely, in Example 3.4.9, we construct an example of a weight function to which no metric is 1-adapted but some metric is 2-adapted.

By (2.3.5), a metric $d \in \mathcal{D}(X, \mathcal{O})$ is M-adapted to a weight function g if and only if there exist $c_1, c_2 > 0$ such that

$$c_1 \delta_M^g(x, y) \leq d(x, y) \leq c_2 \delta_M^g(x, y) \tag{2.4.1}$$

for any $x, y \in X$. By this equivalence, we may think of a metric adapted to a weight function as a "visual metric" associated with the weight function.

If a metric d is M-adapted to a weight function g, then we think of the virtual balls $U_M^g(x, s)$ as the real balls associated with the metric d.

Example 2.4.2 (Figure 2.3) Let us consider the case of the Sierpinski carpet introduced in Example 2.2.8. In this case, the corresponding tree is $(T^{(8)}, \mathcal{A}^{(8)}, \phi)$. Write $T = T^{(8)}$. Define $\eta : T \to (0, 1]$ by $\eta(w) = \frac{1}{3^m}$ for any $w \in (T)_m$. Then η is a weight function and $\Lambda_s^\eta = (T)_m$ if and only if $\frac{1}{3^{m-1}} > s \geq \frac{1}{3^m}$. Let d_* be the (restriction of) Euclidean metric. Then d_* is 1-adapted to h. This can be deduced from the following two observations. First, if $w, v \in (T)_m$ and $K_w \cap K_v \neq \emptyset$, then $\sup_{x \in K_w, y \in K_v} d_*(x, y) \leq \frac{2\sqrt{2}}{3^m}$. Second, if $w, v \in (T)_m$ and $K_w \cap K_v = \emptyset$, then $\inf_{x \in K_w, y \in K_v} d_*(x, y) \geq \frac{1}{3^m}$. In fact, these two facts imply that

$$\frac{1}{3} \delta_1^\eta(x, y) \leq d_*(x, y) \leq 2\sqrt{2} \delta_1^\eta(x, y)$$

for any $x, y \in X$. Next we try another weight function g defined as

$$g(i_1 \ldots i_m) = r_{i_1} \cdots r_{i_m}$$

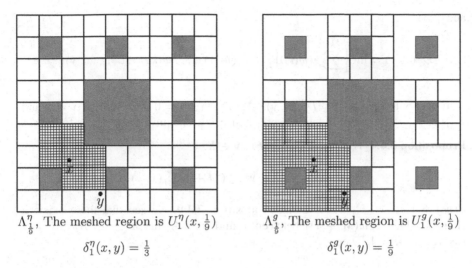

$\Lambda^{\eta}_{\frac{1}{9}}$, The meshed region is $U^{\eta}_1(x, \frac{1}{9})$

$\delta^{\eta}_1(x, y) = \frac{1}{3}$

$\Lambda^{g}_{\frac{1}{9}}$, The meshed region is $U^g_1(x, \frac{1}{9})$

$\delta^g_1(x, y) = \frac{1}{9}$

Fig. 2.3 Visual pre-metrics: the Sierpinski carpet

for any $m \geq 0$ and $i_1, \ldots, i_m \in \{1, \ldots, 8\}$, where

$$r_i = \begin{cases} \frac{1}{9} & \text{if } i \text{ is odd,} \\ \frac{1}{3} & \text{if } i \text{ is even.} \end{cases}$$

Then $\Lambda^g_{\frac{1}{3}} = \{1, \ldots, 8\}$ and

$$\Lambda^g_{\frac{1}{9}} = \{1, 3, 5, 7\} \cup \{i_1 i_2 | i_1 \in \{2, 4, 6, 8\}, i_2 \in \{1, \ldots, 8\}\}.$$

In this case, the existence of an adapted metric is not immediate. However, by [20, Example 1.7.4], it follows that $\eta \underset{\text{GE}}{\sim} g$. (See Definition 3.3.1 for the definition of $\underset{\text{GE}}{\sim}$.) By Theorems 3.5.9 and 2.5.12, there exists a metric $\rho \in \mathcal{D}(X, \mathcal{O})$ that is adapted to g^{α} for some $\alpha > 0$. Furthermore, Theorem 3.6.6 shows that ρ is quasisymmetric to d_*.

There is another "pre-metric" associated with a weight function.

Definition 2.4.3 Let $M \geq 0$. For $x, y \in X$, define

$$D^g_M(x, y) = \inf \left\{ \sum_{i=1}^{k} g(w(i)) \middle| 1 \leq k \leq M + 1, (w(1), \ldots, w(k)) \in \mathcal{CH}_K(x, y) \right\}$$

and

$$D^g(x, y) = \inf\left\{ \sum_{i=1}^k g(w(i)) \,\middle|\, k \geq 1, (w(1), \ldots, w(k)) \in \mathcal{CH}_K(x, y) \right\}.$$

It is easy to see that $0 \leq D_M^g(x, y) \leq 1$, $D_M^g(x, y) = 0$ if and only if $x = y$ and $D_M^g(x, y) = D_M^g(y, x)$. In fact, the pre-metric D_M^g is equivalent to δ_M^g as follows.

Proposition 2.4.4 *For any $M \geq 0$ and $x, y \in X$,*

$$\delta_M^g(x, y) \leq D_M^g(x, y) \leq (M + 1)\delta_M^g(x, y).$$

Proof Set $s_* = \delta_M^g(x, y)$. Using Proposition 2.3.9, we see that there exists a chain $(w(1), \ldots, w(M + 1))$ between x and y such that $w(i) \in \Lambda_{s_*}^g$ for any $i = 1, \ldots, M + 1$. Then

$$D_M^g(x, y) \leq \sum_{i=1}^{M+1} g(w(i)) \leq (M + 1)s_*.$$

Next set $d_* = D_M^g(x, y)$. For any $\epsilon > 0$, there exists a chain $(w(1), \ldots, w(M + 1))$ between x and y such that $\sum_{i=1}^{M+1} g(w(i)) < d_* + \epsilon$. In particular, $g(w(i)) < d_* + \epsilon$ for any $i = 1, \ldots, M + 1$. Hence for any $i = 1, \ldots, M + 1$, there exists $w_*(i) \in \Lambda_{d_*+\epsilon}^g$ such that $K_{w(i)} \subseteq K_{w_*(i)}$. Since $(w_*(1), \ldots, w_*(M + 1))$ is a chain between x and y, it follows that $\delta_M^g(x, y) \leq d_* + \epsilon$. Thus we have shown $\delta_M^g(x, y) \leq D_M^g(x, y)$. □

Combining the above proposition with (2.4.1), we see that d is M-adapted to g if and only if there exist $C_1, C_2 > 0$ such that

$$C_1 D_M^g(x, y) \leq d(x, y) \leq C_2 D_M^g(x, y) \tag{2.4.2}$$

for any $x, y \in X$.

Next we present another condition which is equivalent to a metric being adapted.

Theorem 2.4.5 *Let $g : T \to (0, 1]$ be a weight function and let $M \geq 0$. If $d \in \mathcal{D}(X, \mathcal{O})$, then d is M-adapted to g if and only if the following conditions (ADa) and (ADb)$_M$ hold:*

(ADa) *There exists $c > 0$ such that $\mathrm{diam}(K_w, d) \leq cg(w)$ for any $w \in T$.*

(ADb)$_M$ *For any $x, y \in X$, there exists $(w(1), \ldots, w(k)) \in \mathcal{CH}_K(x, y)$ such that $1 \leq k \leq M + 1$ and*

$$Cd(x, y) \geq \max_{i=1,\ldots,k} g(w(i)),$$

where $C > 0$ is independent of x and y.

Remark In [8, Proposition 8.4], one find an analogous result in the case of partitions associated with expanding Thurston maps. The condition (ADa) and (ADb)$_M$ corresponds their conditions (ii) and (i) respectively.

Proof First assuming (ADa) and (ADb)$_M$, we are going to show (2.4.1). Let $x, y \in X$. By (ADb)$_M$, there exists a chain $(w(1), \dots, w(k))$ between x and y such that $1 \le k \le M+1$ and $Cd(x, y) \ge g(w(i))$ for any $i = 1, \dots, k$. By (G2), there exists $v(i)$ such that $\Sigma_{v(i)} \supseteq \Sigma_{w(i)}$ and $v(i) \in \Lambda^g_{Cd(x,y)}$. Since $(v(1), \dots, v(k))$ is a chain in $\Lambda^g_{Cd(x,y)}$ between x and y, it follows that $Cd(x, y) \ge \delta^g_M(x, y)$.

Next set $t = \delta^g_M(x, y)$. Then there exists a chain $(w(1), \dots, w(M+1)) \in \mathcal{CH}_K(x, y)$ in Λ^g_t. Choose $x_i \in K_{w(i)} \cap K_{w(i+1)}$ for every $i = 1, \dots, M$. Then by (ADa),

$$d(x, y) \le d(x, x_1) + \sum_{i=1}^{M-1} d(x_i, x_{i+1}) + d(x_M, y)$$

$$\le c \sum_{j=1}^{M+1} g(w(i)) \le c(M+1)t = c(M+1)\delta^g_M(x, y).$$

Thus we have (2.4.1).

Conversely, assume that (2.4.1) holds, namely, there exist $c_1, c_2 > 0$ such that $c_1 d(x, y) \le \delta^g_M(x, y) \le c_2 d(x, y)$ for any $x, y \in X$. If $x, y \in K_w$, then $w \in \mathcal{CH}_K(x, y)$. Let $m = \min\{k | g(\pi^k(w)) > g(\pi^{k-1}(w)), k \in \mathbb{N}\}$ and set $s = g(w)$. Then $g(\pi^{m-1}(w)) = s$ and $\pi^{m-1}(w) \in \Lambda^g_s$. Since $\pi^{m-1}(w) \in \mathcal{CH}_K(x, y)$, we have

$$g(w) = s \ge \delta^g_0(x, y) \ge \delta^g_M(x, y) \ge c_1 d(x, y).$$

This immediately yields (ADa).

Set $s_* = c_2 d(x, y)$ for $x, y \in X$. Since $\delta^g_M(x, y) \le c_2 d(x, y)$, there exists a chain $(w(1), \dots, w(M+1))$ in $\Lambda^g_{s_*}$ between x and y. As $g(w(i)) \le s_*$ for any $i = 1, \dots, M+1$, we have (ADb)$_M$. $\qquad \square$

Since (ADa) is independent of M and (ADb)$_M$ implies (ADb)$_N$ for any $N \ge M$, we have the following corollary.

Corollary 2.4.6 *Let $g : T \to (0, 1]$ be a weight function. If $d \in \mathcal{D}(X, \mathcal{O})$ is M-adapted to g for some $M \ge 0$, then it is N-adapted to g for any $N \ge M$.*

Recall that a metric $d \in \mathcal{D}(X, \mathcal{O})$ defines the weight function g_d. So one may ask if d is adapted to the weight function g_d or not. Indeed, we are going to give an example of a metric $d \in \mathcal{D}(X, \mathcal{O})$ which is not adapted to g_d in Example 3.4.8.

Definition 2.4.7 Let $d \in \mathcal{D}(X, \mathcal{O})$. d is said to be adapted if d is adapted to g_d.

Proposition 2.4.8 *Let $d \in \mathcal{D}(X, \mathcal{O})$. d is adapted if and only if there exists a weight function $g : T \to (0, 1]$ to which d is adapted. Moreover, suppose that d is adapted. Then there exists $c_* > 0$ such that*

$$c_* D^d(x, y) \leq d(x, y) \leq D^d(x, y)$$

for any $x, y \in X$, where $D^d = D^{g_d}$.

Proof Necessity direction is immediate. Assume that d is M-adapted to a weight function g. By (ADa) and (ADb)$_M$, for any $x, y \in X$ there exist $k \in \{1, \ldots, M+1\}$ and $(w(1), \ldots, w(k)) \in \mathcal{CH}_K(x, y)$ such that

$$Cd(x, y) \geq \max_{i=1,\ldots,k} g(w(i)) \geq \frac{1}{c} \max_{i=1,\ldots,k} g_d(w(i)).$$

This proves (ADb)$_M$ for the weight function g_d. Since (ADa) for the weight function g_d is immediate, we verify that d is M-adapted to g_d. Now, assuming that d is adapted to g_d, we see

$$c_1 D_M^d(x, y) \leq d(x, y)$$

by (2.4.2). Since $D_M^d(x, y)$ is monotonically decreasing as $M \to \infty$, it follows that

$$c_1 D^d(x, y) \leq d(x, y).$$

On the other hand, if $(w(1), \ldots, w(k)) \in \mathcal{CH}_K(x, y)$, then the triangle inequality yields

$$d(x, y) \leq \sum_{i=1}^{k} g_d(w(i)).$$

Hence $d(x, y) \leq D^d(x, y)$.   □

Let us return to the question on the existence of a metric associated with a given weight function g. As was mentioned at the beginning of this section, we are going to consider a modified/weaker version, i.e. the existence of a metric adapted to g^α for some $\alpha > 0$. Note that if g is a weight function, then so is g^α and $\delta_M^{g^\alpha} = (\delta_M^g)^\alpha$.

To start with, we present a weak version of "triangle inequality" for the family $\{\delta_M^g\}_{M \geq 1}$.

Proposition 2.4.9

$$\delta_{M_1+M_2+1}^g(x, z) \leq \max\{\delta_{M_1}^g(x, y), \delta_{M_2}^g(y, z)\}.$$

Proof Setting $s_* = \max\{\delta_{M_1}^g(x, y), \delta_{M_2}^g(y, z)\}$, we see that there exist a chain $(w(1), \ldots, w(M_1+1))$ between x and y and a chain $(v(1), \ldots, v(M_2+1))$ between y and z such that $w(i), v(j) \in \Lambda_{s_*}^g$ for any i and j. Since $(w(1), \ldots, w(M_1 + 1), v(1), \ldots, v(M_2 + 1))$ is a chain between x and z, we obtain the claim of the proposition. $\qquad\square$

By this proposition, if $\delta_M^g(x, y) \le c\delta_{2M+1}(x, y)$ for any $x, y \in X$, then $\delta_M^g(x, y)$ is so-called quasimetric, i.e.

$$\delta_M^g(x, y) \le c\big(\delta_M^g(x, z) + \delta_M^g(z, y)\big) \qquad (2.4.3)$$

for any $x, y, z \in X$. The coming theorem shows that δ_M^g being a quasimetric is equivalent to the existence of a metric adapted to g^α for some α.

The following definition and proposition give another characterization of the visual pre-metric δ_M^g.

Definition 2.4.10 For $w, v \in T$, the pair (w, v) is said to be m-separated with respect to Λ_s^g if and only if whenever $(w, w(1), \ldots, w(k), v)$ is a chain and $w(i) \in \Lambda_s^g$ for any $i = 1, \ldots, k$, it follows that $k \ge m$.

Proposition 2.4.11 *For any $x, y \in X$ and $M \ge 1$,*

$$\delta_M^g(x, y) = \sup\{s \,|\, (w, v) \text{ is } M\text{-separated with respect to } \Lambda_s^g$$

$$\text{if } w, v \in \Lambda_s^g, x \in K_w \text{ and } y \in K_v\}.$$

A proof of the above proposition is straightforward and left to the readers.

The following theorem gives several equivalent conditions on the existence of a metric adapted to g^α for a given weight function g. In Theorem 2.5.12, those conditions will be shown to be equivalent to the hyperbolicity of the rearrangement of the resolution $(T, \mathcal{B})$ associated with the weight function g and the adapted metric is, in fact, a visual metric in the sense of Gromov. See [12] and [27] for details on visual metric in the sense of Gromov.

Theorem 2.4.12 *Let $M \ge 1$ and let $g \in \mathcal{G}(T)$. The following four conditions are equivalent:*

$(EV)_M$ *There exist $\alpha \in (0, 1]$ and $d \in \mathcal{D}(X, \mathcal{O})$ such that d is M-adapted to g^α.*

$(EV2)_M$ *δ_M^g is a quasimetric, i.e. there exists $c > 0$ such that (2.4.3) holds for any $x, y, z \in X$.*

$(EV3)_M$ *There exists $\gamma \in (0, 1)$ such that $\gamma^n \delta_M^g(x, y) \le \delta_{M+n}^g(x, y)$ for any $x, y \in X$ and $n \ge 1$.*

$(EV4)_M$ *There exists $\gamma \in (0, 1)$ such that $\gamma \delta_M^g(x, y) \le \delta_{M+1}^g(x, y)$ for any $x, y \in X$.*

     Moreover, if $K : T \to \mathcal{C}(X, \mathcal{O})$ is minimal, then all the conditions above are equivalent to the following condition $(EV5)_M$.

$(EV5)_M$ *There exists $\gamma \in (0, 1)$ such that if $(w, v) \in \Lambda_s^g \times \Lambda_s^g$ is M-separated with respect to Λ_s^g, then (w, v) is $(M+1)$-separated with respect to $\Lambda_{\gamma s}^g$.*

The symbol "EV" in the above conditions $(EV)_M$, $(EV1)_M$, ..., $(EV5)_M$ represents "Existence of a Visual metric".

We use the following lemma to prove this theorem.

Lemma 2.4.13 *If there exist* $\gamma \in (0, 1)$ *and* $M \geq 1$ *such that* $\gamma \delta_M^g(x, y) \leq \delta_{M+1}(x, y)$ *for any* $x, y \in X$, *then*

$$\gamma^n \delta_M^g(x, y) \leq \delta_{M+n}^g(x, y)$$

for any $x, y \in X$ *and* $n \geq 1$.

Proof We use an inductive argument. Assume that

$$\gamma^l \delta_M^g(x, y) \leq \delta_{M+l}^g(x, y)$$

for any $x, y \in X$ and $l = 1, \ldots, n$. Suppose $\delta_{M+n+1}^g(x, y) \leq \gamma^{n+1} s$. Then there exists a chain $(w(1), \ldots, w(M + n + 2))$ in $\Lambda_{\gamma^{n+1}s}^g$ between x and y. Choose any $z \in K_{w(M+n+1)} \cap K_{w(M+n+2)}$. Then

$$\gamma^n \delta_M^g(x, z) \leq \delta_{M+n}^g(x, z) \leq \gamma^{n+1} s.$$

Thus we obtain $\delta_M^g(x, z) \leq \gamma s$. Note that $\delta_0^g(z, y) \leq \gamma^{n+1} s$. By Proposition 2.4.9,

$$\gamma \delta_M^g(x, y) \leq \delta_{M+1}^g(x, y) \leq \max\{\delta_M^g(x, z), \delta_0^g(z, y)\} \leq \gamma s.$$

This implies $\delta_M^g(x, y) \leq s$.   □

Proof of Theorem 2.4.12 $(EV)_M \Rightarrow (EV4)_M$: Since d is M-adapted to g^α, by Corollary 2.4.6, d is $(M + 1)$-adapted to g^α as well. By (2.4.1), we obtain $(EV4)_M$. $(EV3)_M \Leftrightarrow (EV4)_M$: This is immediate by Lemma 2.4.13. $(EV3)_M \Rightarrow (EV2)_M$: Let $n = M + 1$. By Proposition 2.4.9, we have

$$c_{2M+1} \delta_M^g(x, y) \leq \delta_{2M+1}^g(x, y) \leq \max\{\delta_M^g(x, z), \delta_M^g(z, y)\} \leq \delta_M^g(x, z) + \delta_M^g(z, y).$$

$(EV2)_M \Rightarrow (EV)_M$: By [16, Proposition 14.5], there exist $c_1, c_2 > 0$, $d \in \mathcal{D}(X, \mathcal{O})$ and $\alpha \in (0, 1]$ such that $c_1 \delta_M^g(x, y)^\alpha \leq d(x, y) \leq c_2 \delta_M^g(x, y)^\alpha$ for any $x, y \in X$. Note that $\delta_M^g(x, y)^\alpha = \delta_M^{g^\alpha}(x, y)$. By (2.4.1), d is M-adapted to g^α. $(EV4)_M \Rightarrow (EV5)_M$: Assume that $w, v \in \Lambda_s^g$. If w and v are not $(M + 1)$-separated with respect to $\Lambda_{\gamma s}^g$, then there exist $w(1), \ldots, w(M) \in \Lambda_{\gamma s}^g$ such that $(w, w(1), \ldots, w(M), v)$ is a chain. Then we can choose $w' \in T_w \cap \Lambda_{\gamma s}^g$ and $v' \in T_v \cap \Lambda_{\gamma s}^g$ so that $(w', w(1), \ldots, w(M), v')$ is a chain. Note that $O_{w'} \neq \emptyset$ and $O_{v'} \neq \emptyset$ since the partition is minimal. Let $x \in O_{w'}$ and let $y \in O_{v'}$. Then $\delta_{M+1}^g(x, y) \leq \gamma s$. Hence by $(EV4)_M$, $\delta_M^g(x, y) \leq s$. There exists a chain $(v(1), v(2), \ldots, v(M + 1))$ in Λ_s^g between x and y. Since $x \in O_{w'} \subseteq O_w$ and $y \in O_{v'} \subseteq O_v$, we see that

$v(0) = w$ and $v(M + 1) = v$. Hence w and v are not M-separated with respect to Λ_s^g.

$(EV5)_M \Rightarrow (EV4)_M$: Assume that $\delta_{M+1}^g(x, y) \leq \gamma s$. Then there exists a chain $(w(1), \ldots, w(M + 2))$ in $\Lambda_{\gamma s}^g$ between x and y. Let w (resp. v) be the unique element in Λ_s^g satisfying $w(1) \in T_w$ (resp. $w(M + 2) \in T_v$). Then (w, v) is not $(M + 1)$-separated in $\Lambda_{\gamma s}^g$. By $(EV5)_M$, (w, v) is not M-separated in Λ_s^g. Hence there exists a chain $(w, v(1), \ldots, v(M - 1), v)$ in Λ_s^g. This implies $\delta_M^g(x, y) \leq s$.

$\square$

2.5   Hyperbolicity of Resolutions and the Existence of Adapted Metrics

In this section, we study the hyperbolicity in Gromov's sense of the resolution $(T, \mathcal{B})$ of a compact metric space X. Roughly speaking the hyperbolicity will be shown to be equivalent to the existence of an adapted metric. More precisely, we define the hyperbolicity of a weight function g as that of certain rearranged subgraph of $(T, \mathcal{B})$ associated with g and show that the weight function g is hyperbolic if and only if there exists a metric adapted to g^α for some $\alpha > 0$. Furthermore, in such a case, the adapted metric is shown to be a "visual metric". See Theorem 2.5.12 for exact statements. Another important point is the "boundary" of the resolution $(T, \mathcal{B})$ is always identified with the original metric space X with or without hyperbolicity of $(T, \mathcal{B})$ as is shown in Theorem 2.5.5. Furthermore, our general framework will be shown to include the preceding results by Elek [15] and Lau-Wang [25] on the constructions of a hyperbolic graph whose hyperbolic boundary belongs to given classes of compact sets.

Throughout this section, $(T, \mathcal{A}, \phi)$ is a locally finite tree with a reference point ϕ, $(X, \mathcal{O})$ is a compact metrizable topological space with no isolated point and $K : T \to \mathcal{C}(X, \mathcal{O})$ is a partition of X parametrized by $(T, \mathcal{A}, \phi)$. Moreover, $(T, \mathcal{B})$ is the resolution of X associated with the partition $K : T \to C(X, \mathcal{O})$.

The first lemma claims that the collection of geodesic rays of $(T, \mathcal{B})$ starting from ϕ equals Σ, which is the collection of geodesic rays of the tree $(T, \mathcal{A})$ starting from ϕ.

Lemma 2.5.1 *If $(w(0), w(1), w(2), \ldots)$ is a geodesic ray from ϕ of $(T, \mathcal{B})$, then $\pi(w(i + 1)) = w(i)$ for any $i = 1, 2, \ldots$. In other word, all the edges of a geodesic ray from ϕ are vertical edges and the collection of geodesic rays of $(T, \mathcal{B})$ coincides with Σ.*

Proof Suppose that $\pi(w(i)) = w(i - 1)$ for any $i = 1, \ldots, n$ and $(w(n), w(n+1))$ is a horizontal edge. Then $|w(i)| = i$ for any $i = 0, 1, \ldots, n$ and $|w(n + 1)| = n$. Since $d_{(T, \mathcal{B})}(\phi, w(n+1)) = n$, the sequence $(\phi, w(1), \ldots, w(n), w(n+1))$ can not be a geodesic. Hence there exists no horizontal edge in $(w(0), w(1), w(2), \ldots)$. $\square$

The second statement of the following proposition is the restatement of Proposition 2.2.12.

Proposition 2.5.2 (= Proposition 2.2.12) *Let $\omega, \tau \in \Sigma$.*

(1) *For any $n \geq 1$,*

$$d_{(T,\mathcal{B})}([\omega]_n, [\tau]_n) \leq d_{(T,\mathcal{B})}([\omega]_{n+1}, [\tau]_{n+1}).$$

(2) $\sigma(\omega) = \sigma(\tau)$ *if and only if*

$$\sup_{n \geq 1} d_{(T,\mathcal{B})}([\omega]_n, [\tau]_n) < +\infty.$$

To prove the above proposition, we need to study the structure of geodesics of $(T, d_{(T,\mathcal{B})})$.

Definition 2.5.3

(1) Let $w, v \in (T)_m$ for some $m \geq 0$. The pair (w, v) is called horizontally minimal if and only if there exists a geodesic of the resolution $(T, \mathcal{B})$ between w and v which consists only of horizontal edges.
(2) Let $w \neq v \in T$. Then a geodesic **b** of $(T, \mathcal{B})$ between w and v is called a bridge if and only if there exist $i, j \geq 0$ and a horizontal geodesic $(v(1), \ldots, v(k))$ such that $\pi^i(w) = v(1), \pi^j(v) = v(k)$ and

$$\mathbf{b} = (w, \pi(w), \ldots, \pi^i(w), v(2), \ldots, v(k-1), \pi^j(v), \ldots, \pi(v), v).$$

The number $|v(1)|$ is called the height of the bridge. Also $(w, \pi(w), \ldots, \pi^i(w))$, $(v(1), \ldots, v(k))$ and $(\pi^j(v), \ldots, \pi(v), v)$ are called the ascending part, the horizontal part and the descending part respectively.

Lemma 2.5.4 *For any $w, v \in T$, there exists a bridge between w and v.*

Proof Let $(w(1), \ldots, w(m))$ be a geodesic of $(T, \mathcal{B})$ between w and v. Note that there exists no dent, which is a segment $(w(i), \ldots, w(k), w(k+1))$ satisfying $|w(i+1)| = |w(i)|+1, |w(i+1)| = |w(i+2)| = \ldots = |w(k)|$ and $|w(k+1)| = |w(k)|-1$, because $(w(i), \pi(w(i+2)), \ldots, \pi(w(k-1)), w(k+1))$ is a geodesic with shorter length. Therefore, if $m_* = \min\{|w(i)| : i = 1, \ldots, m\}$, then there exist $i_* < j_*$ such that $\{i : |w(i)| = m_*\} = \{i : i_* \leq i \leq j_*\}$ and $|w(i)|$ is monotonically nonincreasing on $I_1 = \{i | 1 \leq i \leq i_*\}$ and monotonically nondecreasing on $I_2 = \{i | j_* \leq i \leq m\}$. Let

$$\mathcal{A}_d = \{i | 1 \leq i \leq i_* - 1, \pi(w(i)) = w(i+1)\}.$$

Since $|w(i)|$ is monotonically nonincreasing, $\#\mathcal{A}_d = |w| - m_*$. Set $v(i) = \pi^{|w(i)|-m_*}(w(i))$. Then $(v(1), \ldots, v(i_*))$ is a horizontal chain in $(T)_{m_*}$. Note that $v(i) = v(i+1)$ for $i \in \mathcal{A}_d$. Removing such redundant parts, we obtain a subchain

$(\omega(1), \ldots, \omega(i_* - |w| + m_*))$ of $(v(1), \ldots, v(i_*))$. Set $\tilde{w}(i) = \pi^i(w)$. Note that $\tilde{w}(|w| - m_*) = \omega(1)$. Replacing $(w(1), \ldots, w(i_*))$ by $(\tilde{w}(0), \ldots, \tilde{w}(|w| - m_*), \omega(2), \ldots, \omega(i_* - |w| + m_*))$, we obtain a geodesic between w to v. Operating the similar procedure on $(w(j_*), \ldots, w(m))$, we have a bridge between w and v.

□

Proof of Proposition 2.5.2 Set $\omega_m = [\omega]_m$ and $\tau_m = [\tau]_m$ for any $m \geq 1$. Let $\mathbf{b} = (\omega_m, \ldots, \omega_{m-n}, \ldots, \tau_{m-n}, \ldots, \tau_m)$ be a bridge between ω_m and τ_m, where $(\omega_{m-n}, \ldots, \tau_{m-n})$ is the horizontal part.

(1) If $n = 0$, then $\pi(\mathbf{b})$ is a horizontal chain in $(T)_{m-1}$ between ω_{m-1} and τ_{m-1}. Hence

$$d_{(T,\mathcal{B})}(\omega_{m-1}, \tau_{m-1}) \leq d_{(T,\mathcal{B})}(\omega_m, \tau_m).$$

If $n \geq 1$, then $(\omega_{m-1}, \ldots, \omega_{m-n}, \ldots, \tau_{m-n}, \ldots, \tau_{m-1})$ is a bridge between ω_{m-1} and τ_{m-1} and hence

$$d_{(T,\mathcal{B})}(\omega_{m-1}, \tau_{m-1}) + 2 = d_{(T,\mathcal{B})}(\omega_m, \tau_m).$$

(2) Assume that $\sup_{n \geq 1} d_{(T,\mathcal{B})}(\omega_n, \tau_n) = N < +\infty$. Let $d \in \mathcal{D}(X, \mathcal{O})$. Then g_d is a weight function by Proposition 2.3.5 and so $\max_{w \in (T)_m} \operatorname{diam}(K_w, d) \to 0$ as $m \to \infty$. Set $x = \sigma(\omega)$ and $y = \sigma(\tau)$. Since $n \leq d_{(T,\mathcal{B})}(\omega_m, \tau_m) \leq N$ and the length of $(\omega_{m-n}, \ldots, \tau_{m-n})$ is at most N, it follows that $d(x, y) \leq N \max_{w \in (T)_{m-N}} \operatorname{diam}(K_w, d) \to 0$ as $m \to \infty$. Therefore $x = y$. Conversely, if $x = y$, then $x = y \in K_{[\omega]_m} \cap K_{[\tau]_m}$ for any $m \geq 0$. Therefore $d_{(T,\mathcal{B})}([\omega]_m, [\tau]_m) \leq 1$ for any $m \geq 0$.

□

Proposition 2.5.2 immediately yields the following theorem.

Theorem 2.5.5 *Define an equivalence relation $\sim$ on the collection Σ of the geodesic rays as $\omega \sim \tau$ if and only if $\sup_{n \geq 1} d_{(T,\mathcal{B})}([\omega]_n, [\tau]_n) < +\infty$. Let $\mathcal{O}_*$ be the natural quotient topology of $\Sigma / \sim$ induced by the metric ρ_* on Σ. Then*

$$(\Sigma/\sim, \mathcal{O}_*) = (X, \mathcal{O}),$$

where we identify $\Sigma / \sim$ with X through the map $\sigma : \Sigma \to X$.

For a Gromov hyperbolic graph, the quotient of the collection of geodesic rays by the above equivalence relation $\sim$ is called the hyperbolic boundary of the graph. In our framework, however, due to Theorem 2.5.5, $\Sigma / \sim$ can be always identified with X even if $(T, \mathcal{B})$ is not hyperbolic.

Next, we introduce the notion of (Gromov) hyperbolicity of $(T, \mathcal{B})$. Here we give only basic accounts needed in our work. See [12, 27] and [29] for details of the general framework of Gromov hyperbolic metric spaces.

Definition 2.5.6 Define the Gromov product of $w, v \in T$ in $(T, \mathcal{B})$ with respect to ϕ as

$$(w|v)_{(T,\mathcal{B}),\phi} = \frac{d_{(T,\mathcal{B})}(\phi, w) + d_{(T,\mathcal{B})}(\phi, v) - d_{(T,\mathcal{B})}(w, v)}{2}.$$

The graph $(T, \mathcal{B})$ is called η-hyperbolic (in the sense of Gromov) if and only if

$$(w|v)_{(T,\mathcal{B}),\phi} \geq \min\{(w|u)_{(T,\mathcal{B}),\phi}, (u|v)_{(T,\mathcal{B}),\phi}\} - \eta$$

for any $w, v, u \in T$. $(T, \mathcal{B})$ is called hyperbolic if and only if it is η-hyperbolic for some $\eta \in \mathbb{R}$.

It is known that the hyperbolicity can be defined by the thinness of geodesic triangles.

Definition 2.5.7 We say that all the geodesic triangles in $(T, \mathcal{B})$ are δ-thin if and only if for any $u, v, w \in T$, if $\mathbf{b}(a, b)$ is geodesic between a and b for each $(a, b) \in \{(u, v), (v, w), (w, u)\}$, then $\mathbf{b}(v, w)$ is contained in δ-neighborhood of $\mathbf{b}(u, v) \cup \mathbf{b}(w, u)$ with respect to $d_{(T,\mathcal{B})}$.

The following theorem is one of the basic facts in the theory of Gromov hyperbolic spaces. A proof can be seen in [29] for example.

Theorem 2.5.8 $(T, \mathcal{B})$ is η-hyperbolic for some $\eta > 0$ if and only if all the geodesic triangles in $(T, \mathcal{B})$ are δ-thin for some $\delta > 0$.

The next theorem gives a criterion of the hyperbolicity of the resolution $(T, \mathcal{B})$. It has explicitly stated and proven by Lau and Wang in [25]. However, Elek had already used essentially the same idea in [15] to construct a hyperbolic graph which is quasi-isometric to the hyperbolic cone of a compact metric space. In fact, we are going to recover their works as a part of our general framework later in this section.

Theorem 2.5.9 The resolution $(T, \mathcal{B})$ of X is hyperbolic if and only if there exists $L \geq 1$ such that

$$d_{(T,\mathcal{B})}(w, v) \leq L \tag{2.5.1}$$

for any horizontally minimal pair $(w, v) \in \cup_{m \geq 1}((T)_m \times (T)_m)$.

Remark As is shown in the proof, if all the geodesic triangles in $(T, \mathcal{B})$ are δ-thin, then L can be chosen as $4\delta + 1$. Conversely, if (2.5.1) is satisfied, then $(T, \mathcal{B})$ is $\frac{3}{2}L$-hyperbolic.

Since our terminologies and notations differ much from those in [15] and [25], we are going to present a proof of Theorem 2.5.9 for reader's sake.

Proof Assume that all the geodesic triangles in $(T, \mathcal{B})$ are δ-thin. Let $(w, v) \in (T)_m$ be horizontally minimal. Consider the geodesic triangle consists of $p_1 =$

$(w, \pi(w), \ldots, \pi^m(w))$, which is the vertical geodesic between w and ϕ, $p_2 = (v, \pi(v), \ldots, \pi^m(v))$, which is the vertical geodesic between v and ϕ, and $p_3 = (u(1), \ldots, u(k+1))$, which is the horizontal geodesic between w and v. Since all the geodesic triangles in $(T, \mathcal{B})$ are δ-thin, for any i, either there exists $w' \in p_1$ such that $d_{(T,\mathcal{B})}(w', u(i)) \leq \delta$ or there exists $v' \in p_2$ such that $d_{(T,\mathcal{B})}(v', u(i)) \leq \delta$. Suppose that the former is the case. Since $d_{(T,\mathcal{B})}(w, w') = |w| - |w'|$ is the smallest steps from the level $|w| = |u(i)|$ to $|w'|$, it follows that $d_{(T,\mathcal{B})}(w, w') \leq d_{(T,\mathcal{B})}(w', u(i))$. Hence

$$d_{(T,\mathcal{B})}(w, u(i)) \leq d_{(T,\mathcal{B})}(w, w') + d_{(T,\mathcal{B})}(w', u(i)) \leq 2d_{(T,\mathcal{B})}(w', u(i)) \leq 2\delta.$$

Considering the latter case as well, we conclude that either $d_{(T,\mathcal{B})}(w, u(i))) \leq 2\delta$ or $d_{(T,\mathcal{B})}(v, u(i)) \leq 2\delta$ for any i. This shows that $d_{(T,\mathcal{B})}(w, v) \leq 4\delta + 1$.

Conversely, assume (2.5.1). For ease of notation, we use $(w|v)$ in place of $(w|v)_{(T,\mathcal{B}),\phi}$, We are going to show that $(T, \mathcal{B})$ is $\frac{3}{2}L$-hyperbolic, namely,

$$(w(1)|w(2)) \geq \min\{(w(2)|w(3)), (w(3)|w(1))\} - \frac{3}{2}L \qquad (2.5.2)$$

for any $w(1), w(2), w(3) \in T$. For $(i, j) \in \{(1, 2), (2, 3), (3, 1)\}$, let $\mathbf{b}_{ij}$ be a bridge between $w(i)$ and $w(j)$ and let m_{ij}, l_{ij} and m_{ji} be the lengths of the ascending part, the horizontal part and the descending part respectively. Also set h_{ij} be the height of the bridge $\mathbf{b}_{ij}$. Then

$$(w(i)|w(j)) = \frac{h_{ij} + m_{ij} + h_{ij} + m_{ji} - (m_{ij} + m_{ji} + l_{ij})}{2} = h_{ij} - \frac{l_{ij}}{2}.$$

Without loss of generality, we may assume that $h_{23} \geq h_{31}$. Then we have three cases;

Case 1: $h_{12} \geq h_{23} \geq h_{31}$,
Case 2: $h_{23} \geq h_{12} \geq h_{31}$,
Case 3: $h_{23} \geq h_{31} \geq h_{12}$.

In Case 1 and Case 2, since $h_{31} - h_{12} \leq 0$ and $l_{12} \leq L$, it follows that

$$(w(3)|w(1)) - (w(1)|w(2)) = h_{31} - h_{12} + \frac{l_{12}}{2} - \frac{l_{31}}{2} \leq \frac{L}{2}.$$

Thus (2.5.2) holds. In Case 3, let $v(1) \in (T)_{h_{31}}$ belong to the ascending part of $\mathbf{b}_{12}$ and let $v(2) \in (T)_{h_{31}}$ belong to the descending part of $\mathbf{b}_{12}$. Moreover, let $\mathbf{b}_{31}^h$ and $\mathbf{b}_{23}^h$ be the horizontal parts of $\mathbf{b}_{31}$ and $\mathbf{b}_{23}$ respectively. Then the combination of $\mathbf{b}_{31}^h$ and $\pi^{h_{23}-h_{31}}(\mathbf{b}_{23}^h)$ gives a chain between $v(1)$ and $v(2)$ whose length is no greater than $l_{31} + l_{23}$. Since the segment of $\mathbf{b}_{12}$ connecting $v(1)$ and $v(2)$ is a geodesic, we have

$$2(h_{31} - h_{12}) + l_{12} \leq l_{31} + l_{23} \leq 2L.$$

Therefore, it follows that $h_{31} - h_{12} \leq L$. This implies

$$(w(3)|w(1)) - (w(1)|w(2)) = h_{31} - h_{12} + \frac{l_{12}}{2} - \frac{l_{31}}{2} \leq L + \frac{L}{2} = \frac{3}{2}L.$$

Thus we have obtained (2.5.2) in this case as well.

Note that so far weight functions play no role in the statements of results in this section. In order to take weight functions into account, we are going to introduce the rearranged resolution $(\widetilde{T}^{g,r}, \mathcal{B}_{\widetilde{T}^{g,r}})$ associated with a weight function g and give the definition of hyperbolicity of the weight function g in terms of the rearranged resolution.

Definition 2.5.10 Let $g \in \mathcal{G}(T)$ and let $r \in (0, 1)$. For $m \geq 0$, define $(\widetilde{T}^{g,r})_m = \Lambda_{r^m}^g$ and

$$\widetilde{T}^{g,r} = \bigcup_{m \geq 0} (\widetilde{T}^{g,r})_m.$$

$\widetilde{T}^{g,r}$ is naturally equipped with a tree structure inherited from T. Define $K_{\widetilde{T}^{g,r}} : \widetilde{T}^{g,r} \to \mathcal{C}(X, \mathcal{O})$ by $K_{\widetilde{T}^{g,r}} = K|_{\widetilde{T}^{g,r}}$. The collection of geodesic rays of the tree $\widetilde{T}^{g,r}$ starting from ϕ is denoted by $\Sigma_{\widetilde{T}^{g,r}}$. Define $\sigma_{\widetilde{T}^{g,r}} : \Sigma_{\widetilde{T}^{g,r}} \to X$ by $\sigma_{\widetilde{T}^{g,r}}(\omega) = \cap_{m \geq 0} K_{\omega(m)}$ for any $\omega = (\phi, \omega(1), \ldots) \in \Sigma_{\widetilde{T}^{g,r}}$. For any $w \in \Lambda_{r^{m+1}}^g$, the unique $v \in \Lambda_{r^m}^g$ satisfying $w \in T_v$ is denoted by $\pi^{g,r}(w)$. Also we set $S^{g,r}(w) = \{v | v \in \Lambda_{r^{m+1}}^d, v \in T_w\}$ for $w \in \Lambda_{r^m}^g$. Define the horizontal edges of $\widetilde{T}^{g,r}$ as

$$E_{g,r}^h = \bigcup_{n \geq 1} \{(w, v) | w, v \in (\widetilde{T}^{g,r})_n, w \neq v, K_w \cap K_v \neq \emptyset\}.$$

Moreover, we define the totality of the horizontal and vertical edges $\mathcal{B}_{\widetilde{T}^{g,r}}$ by

$$\mathcal{B}_{\widetilde{T}^{g,r}} = \{(w, v) | (w, v) \in E_{g,r}^h \text{ or } w = \pi^{g,r}(v) \text{ or } v = \pi^{g,r}(w)\}.$$

The graph $(\widetilde{T}^{g,r}, \mathcal{B}_{\widetilde{T}^{g,r}})$ is called the rearranged resolution of X associated with a weight function g.

Remark Even if $m \neq n$, it may happen that $\Lambda_{r^m}^g \cap \Lambda_{r^n}^g \neq \emptyset$. In such a case, for $w \in \Lambda_{r^m}^g \cap \Lambda_{r^n}^g$, we regard $w \in (\widetilde{T}^{g,r})_m$ and $w \in (\widetilde{T}^{g,r})_n$ as different elements in $\widetilde{T}^{g,r}$. More precisely, the exact definition of $\widetilde{T}^{g,r}$ should be $\widetilde{T}^{g,r} = \cup_{m \geq 0}(\{m\} \times \Lambda_{r^m}^g)$ and the associated partition $K_{\widetilde{T}^{g,r}} : \widetilde{T}^{g,r} \to \mathcal{C}(X, \mathcal{O})$ is defined as $K_{\widetilde{T}^{g,r}}((m, w)) = K_w$.

Remark $\Sigma_{\widetilde{T}^{g,r}}$ and $\sigma_{\widetilde{T}^{g,r}}$ can be naturally identified with Σ and σ respectively.

Definition 2.5.11 A weight function g is said to be hyperbolic if and only if the rearranged resolution $(\widetilde{T}^{g,r}, \mathcal{B}_{\widetilde{T}^{g,r}})$ is hyperbolic for some $r \in (0, 1)$.

The next theorem shows that the hyperbolicity of a weight function g is equivalent to the existence of a "visual metric" associated with g. It also implies that the quantifier "for some $r \in (0, 1)$" in Definition 2.5.11 can be replaced by "for any $r \in (0, 1)$".

Theorem 2.5.12 *Let g be a weight function. Then the following three conditions are equivalent:*

(1) *There exists $M \geq 1$ such that $(EV)_M$ is satisfied, i.e. there exist $d \in \mathcal{D}(X, \mathcal{O})$ and $\alpha > 0$ such that d is M-adapted to g^α.*
(2) *The weight function g is hyperbolic.*
(3) *$(\widetilde{T}^{g,r}, \mathcal{B}_{\widetilde{T}^{g,r}})$ is hyperbolic for any $r \in (0, 1)$.*

Moreover, if any of the above conditions is satisfied, then there exist $c_1, c_2 > 0$ such that

$$c_1 \delta_M^g(x, y) \leq r^{(x|y)_{\widetilde{T}^{g,r}}} \leq c_2 \delta_M^g(x, y) \tag{2.5.3}$$

for any $x, y \in X$, where

$$(x|y)_{\widetilde{T}^{g,r}} = \sup \left\{ \lim_{n,m \to \infty} (\omega(n)|\tau(m))_{(\widetilde{T}^{g,r}, \mathcal{B}_{\widetilde{T}^{g,r}}), \phi} \right|$$

$$\omega = (\phi, \omega(1), \dots), \tau = (\phi, \tau(1), \dots) \in \Sigma_{\widetilde{T}^{g,r}}, \sigma_{\widetilde{T}^{g,r}}(\omega) = x, \sigma_{\widetilde{T}^{g,r}}(\tau) = y \right\}.$$

Remark The proof of Theorem 2.5.12 shows that if every geodesics triangle of $(\widetilde{T}^{g,r}, \mathcal{B}_{\widetilde{T}^{g,r}})$ is η-thin, then $(EV)_M$ is satisfied for $M = \min\{m | m \in \mathbb{N}, 4\eta+1 \leq m\}$.

By (2.5.3), if d is M-adapted to g^α, then there exist $c_1, c_2 > 0$ such that

$$c_1 d(x, y) \leq (r^\alpha)^{(x|y)_{\widetilde{T}}} \leq c_2 d(x, y)$$

for any $x, y \in X$. Then the metric d is called a visual metric on the hyperbolic boundary X of $(\widetilde{T}^{g,r}, \mathcal{B}_{\widetilde{T}^{g,r}})$ in the framework of Gromov hyperbolic metric spaces. See [12] and [27] for example.

About the original resolution $(T, \mathcal{B})$, we have the following corollary.

Corollary 2.5.13 *For $r \in (0, 1)$, define a weight function h_r by $h_r(w) = r^{-|w|}$ for $w \in T$. Then $(T, \mathcal{B})$ is hyperbolic if and only if there exist a metric $d \in \mathcal{D}(X, \mathcal{O})$, $M \geq 1$ and $r \in (0, 1)$ such that d is M-adapted to h_r.*

To show the hyperbolicity of a weight function g, the existence of an adapted metric is (an equivalent condition as we have seen in Theorem 2.5.12 but) too restrictive in some cases. In fact, the notion of "weakly adapted" metric is often more useful as we will see in Example 2.5.18 and 2.5.19.

Definition 2.5.14 Let $d \in \mathcal{D}(X, \mathcal{O})$. For $r \in (0, 1]$, $s > 0$ and $x \in X$, define

$$\widetilde{B}^r_{d,g}(x, s) = \{y \mid y \in B_d(x, s), \text{ there exists a horizontal chain } (w(1), \ldots, w(k))$$

in Λ^g_r between x and y such that $K_{w(i)} \cap B_d(x, s) \neq \emptyset$ for any $i = 1, \ldots, k\}$.

A metric $d \in \mathcal{D}(X, \mathcal{O})$ is said to be weakly M-adapted to a weight function g if and only if there exist $c_1, c_2 > 0$ such that

$$\widetilde{B}^r_{d,g}(x, c_1 r) \subseteq U^g_M(x, r) \subseteq B_d(x, c_2 r)$$

for any $x \in X$ and $r \in (0, 1]$.

Since $\widetilde{B}^r_{d,g}(x, cr) \subseteq B_d(x, cr)$, we immediately have the following fact.

Proposition 2.5.15 *If $d \in \mathcal{D}(X, \mathcal{O})$ is M-adapted to a weight function g, then it is weakly M-adapted to g.*

The next proposition gives a sufficient condition for a metric being weakly adapted, which will be applied in Examples 2.5.18 and 2.5.19.

Proposition 2.5.16 *If there exist $c_1, c_2 > 0$ and $M \in \mathbb{N}$ such that*

$$\text{diam}(K_w, d) \leq c_1 r \tag{2.5.4}$$

for any $w \in \Lambda^g_r$ and

$$\#(\{w \mid w \in \Lambda^g_r, B_d(x, c_2 r) \cap K_w \neq \emptyset\}) \leq M + 1 \tag{2.5.5}$$

for any $x \in X$ and $r \in (0, 1]$, then d is weakly M-adapted to g.

Proof Assume that $y \in U^g_M(x, r)$. Then there exists a chain $(w(1), \ldots, w(M+1))$ between x and y in Λ^g_r. Choose $x_i \in K_{w(i)} \cap K_{w(i+1)}$ for any i. Set $x_0 = x$ and $x_{M+1} = y$ Then by (2.5.4), it follows that

$$d(x, y) \leq \sum_{i=0}^{M} d(x_i, x_{i+1}) \leq (M+1)c_1 r.$$

Hence $y \in B_d(x, (M+2)c_1 r)$. This implies $U^g_M(x, r) \subseteq B_d(x, (M+2)c_1 r)$.

Next, let $y \in \widetilde{B}^r_d(x, c_2 r)$. Then there exists a chain $(w(1), \ldots, w(k))$ between x and y in Λ^g_r such that $K_{w(i)} \cap B_d(x, c_2 r) \neq \emptyset$ for any $i = 1, \ldots, k$. We may assume that $w(i) \neq w(j)$ if $i \neq j$. Then by (2.5.5), we see that $k \leq M + 1$. This yields that $y \in U^g_M(x, r)$. Thus we have shown that d is weakly M-adapted to g. $\square$

Proposition 2.5.17 *Let g be a weight function. If there exists a metric $d \in \mathcal{D}(X, \mathcal{O})$ that is weakly M-adapted to g^α for some $M \geq 1$ and $\alpha > 0$, then $(\widetilde{T}^{g,r}, \mathcal{B}_{\widetilde{T}^{g,r}})$ is hyperbolic for any $r \in (0, 1]$.*

There have been several works on the construction of a hyperbolic graph whose hyperbolic boundary coincides with a given compact metric space. For example, Elek[15] has studied the case for an arbitrary compact subset of $\mathbb{R}^n$, and Lau-Wang[25] has considered self-similar sets satisfying the open set condition. Due to the above proposition, we may integrate these works into our framework. See Example 2.5.18 and 2.5.19 for details.

For ease of notations, we use $\widetilde{T}$, $\widetilde{\pi}$, $\Sigma_{\widetilde{T}}$, $\sigma_{\widetilde{T}}$ and $\mathcal{B}_{\widetilde{T}}$ to denote $\widetilde{T}^{g,r}$, $\pi^{g,r}$, $\Sigma_{\widetilde{T}^{g,r}}$, $\sigma_{\widetilde{T}^{g,r}}$ and $\mathcal{B}_{\widetilde{T}^{g,r}}$, respectively. Moreover, we write $d_{\widetilde{T}} = d_{(\widetilde{T},\mathcal{B}_{\widetilde{T}})}$, which is the geodesic metric of $(\widetilde{T}, \mathcal{B}_{\widetilde{T}})$.

Proof Assume that there exist a metric $d \in \mathcal{D}(X, \mathcal{O})$, $M \geq 1$ and $\alpha > 0$ such that d is weakly M-adapted to g^α. Then there exist $c_1, c_2 > 0$ such that

$$\widetilde{B}^s_{d,g^\alpha}(x, c_1 s) \subseteq U^{g^\alpha}_M(x, s) \subseteq B_d(x, c_2 s)$$

for any $x \in X$ and $s \in (0, 1]$. In particular, $d(x, y) \leq c_2 \delta^{g^\alpha}_M(x, y)$ for any $x, y \in X$. Fix $r \in (0, 1)$. Suppose that $w, v \in (\widetilde{T})_n$, $x_1 \in K_w$, $x_2 \in K_v$ and $K_w \cap K_v \neq \emptyset$. Since $\delta^g_1(x_1, x_2) \leq r^n$, it follows that

$$d(x_1, x_2) \leq c_2 r^{\alpha n}. \tag{2.5.6}$$

Let $m \geq 1$. Suppose that $w_*, v_* \in (\widetilde{T})_n$, (w_*, v_*) is horizontally minimal and $3m - 2 \leq d_{\widetilde{T}}(w_*, v_*) \leq 3m$. Let $(w(1), \ldots, w(3m + 1))$ be the horizontal chain of $(\widetilde{T}, \mathcal{B}_{\widetilde{T}})$ between w_* and v_*. Let $x \in K_{w_*}$ and let $y \in K_{v_*}$. By (2.5.6) for any $z \in \cup^{3m+1}_{i=1} K_{w_*}$,

$$d(x, z) \leq 3m c_2 r^{\alpha n}.$$

Choose m sufficiently large so that $3m c_2 r^{\alpha m} < c_1$. Then

$$d(x, z) < c_1 r^{\alpha(n-m)}.$$

Set $v(i) = \widetilde{\pi}^m(w(i))$ for $i = 1, \ldots, 3m + 1$, then $(v(1), \ldots, v(3m + 1))$ is a horizontal chain in $(\widetilde{T})_{n-m}$ between x and y and $K_{v(i)} \cap B_d(x, c_1 r^{\alpha(n-m)}) \supseteq K_{w(i)} \cap B_d(x, c_1 r^{\alpha(n-m)}) \neq \emptyset$. Since $(\widetilde{T})_{n-m} = \Lambda^{g^\alpha}_{r^{\alpha(n-m)}}$, we have

$$y \in \widetilde{B}^{r^{\alpha(n-m)}}_{d,g^\alpha}(x, c_1 r^{\alpha(n-m)}) \subseteq U^{g^\alpha}_M(x, r^{\alpha(n-m)}) = U^g_M(x, r^{n-m}).$$

So there exists a horizontal chain $(u(1), \ldots, u(M + 1))$ in $(\widetilde{T})_{n-m}$ between x and y. Combining this horizontal chain with vertical geodesics $(w_*, \ldots, \widetilde{\pi}^m(w_*))$ and $(\widetilde{\pi}^m(v_*), \ldots, v_*)$, we have a chain of $(\widetilde{T}, \mathcal{B}_{\widetilde{T}})$ between w_* and v_* whose length is $M + 2 + 2m$. Therefore,

$$M + 2 + 2m \geq d_{\widetilde{T}}(w_*, v_*) \geq 3m - 2.$$

Hence $M + 4 \geq m$. Applying Theorem 2.5.9 to $(\widetilde{T}, \mathcal{B}_{\widetilde{T}})$, we verify that $(\widetilde{T}, \mathcal{B}_{\widetilde{T}})$ is hyperbolic.   □

Proof of Theorem 2.5.12 (1) $\Rightarrow$ (3) Proposition 2.5.17 suffices.

(3) $\Rightarrow$ (2) This is immediate.

(2) $\Rightarrow$ (1) Assume that all the geodesic triangles in $(\widetilde{T}, \mathcal{B}_{\widetilde{T}})$ are δ-thin. Set $L = \min\{m | m \in \mathbb{N}, 4\delta + 1 \leq m\}$. Then Theorem 2.5.9 shows that the length of every horizontal geodesic is no greater than L. For ease of notation, we use $(w|v)_{\widetilde{T}}$ to denote the Gromov product of w and v in $(\widetilde{T}, \mathcal{B}_{\widetilde{T}})$ with respect to ϕ. Let $x \neq y \in X$ and let $\omega = (\phi, \omega(1), \ldots), \tau = (\phi, \tau(1), \ldots) \in \Sigma_{\widetilde{T}}$ satisfy $\sigma_{\widetilde{T}}(\omega) = x$ and $\sigma_{\widetilde{T}}(\tau) = y$. Applying Proposition 2.5.2 to $\widetilde{T}$, we see that there exists $m_* \in \mathbb{N}$ such that $d_{\widetilde{T}}(\omega(m), \tau(m)) > L$ for any $m \geq m_*$. Let **b** be a bridge between $\omega(m_*)$ and $\tau(m_*)$. If k_* is the height of **b** and $(\omega(k_*), w(1), \ldots, w(l-1), \tau(k_*))$ is the horizontal part of **b**, then **b** is the concatenation of $(\omega(m_*), \ldots, \omega(k_*))$, $(\omega(k_*), w(1), \ldots, w(l-1), \tau(k_*))$ and $(\tau(k_*), \ldots, \tau(m_*))$. If $m, n \geq m_*$, then $(\omega(m), \ldots, \omega(k_*), w(1), \ldots, w(l-1), \tau(k_*), \ldots, \tau(n))$ is a bridge between $\omega(m)$ and $\tau(n)$. Therefore,

$$(\omega(m)|\tau(n))_{\widetilde{T}} = k_* - \frac{l}{2}.$$

Hence, if we define

$$(\omega|\tau)_{\widetilde{T}} = \lim_{m,n \to \infty} (\omega(m)|\tau(n))_{\widetilde{T}},$$

then $(\omega|\tau)_{\widetilde{T}} = k_* - \frac{l}{2}$.

Applying Theorem 2.5.9 to $(\widetilde{T}, \mathcal{B}_{\widetilde{T}})$, we see that the length of any horizontal geodesic is no greater than L. Since the length of the horizontal part of **b** is l, it follows that $l \leq L$. Therefore letting $s_* = \delta_L^g(x, y)$, then we see that

$$s_* \leq r^{k_*} \leq r^{k_* - \frac{l}{2}} = r^{(\omega|\tau)_{\widetilde{T}}}. \tag{2.5.7}$$

Choose n_* so that $r^{k_* + n_*} \geq s_* > r^{k_* + n_* + 1}$. Then there exists a horizontal chain $(v(1), \ldots, v(L+1))$ in $(\widetilde{T})_{k_* + n_*}$ such that $x \in K_{v(1)}$ and $y \in K_{v(L+1)}$. Hence $(\omega(k_* + n_*), v(1), \ldots, v(L+1), \tau(k_* + n_*))$ is a chain between $\omega(k_* + n_*)$ and $\tau(k_* + n_*)$. Comparing this chain with $(\omega(k_* + n_*), \ldots, \omega(k_*), w(1), \ldots, w(l-1), \tau(k_*), \ldots, \tau(k_* + n_*))$, we obtain

$$2n_* + l \leq L + 2.$$

This implies $k_* + n_* + 1 - \frac{L}{2} - 2 \le k_* - \frac{l}{2}$. Therefore

$$r^{(\omega|\tau)_{\tilde{T}}} = r^{k_* - \frac{l}{2}} \le r^{k_* + n_* + 1} r^{-\frac{L}{2} - 2} \le r^{-\frac{L}{2} - 2} s_*. \tag{2.5.8}$$

Set $c_1 = 1$ and $c_2 = r^{-\frac{L}{2} - 2}$. Then we have

$$c_1 \delta_L^g(x, y) \le r^{(\omega|\tau)_{\tilde{T}}} \le c_2 \delta_L^g(x, y).$$

Define $(x|y)_{\tilde{T}} = \sup\{(\omega|\tau)_{\tilde{T}} | \omega, \tau \in \Sigma_{\tilde{T}}, \sigma_{\tilde{T}}(\omega) = x, \sigma_{\tilde{T}}(\tau) = y\}$. Then

$$c_1 \delta_L^g(x, y) \le r^{(x|y)_{\tilde{T}}} \le c_2 \delta_L^g(x, y). \tag{2.5.9}$$

It is known that if $(\tilde{T}, \mathcal{B}_{\tilde{T}})$ is hyperbolic, then $r^{(x|y)_{\tilde{T}}}$ is a quasimetric. Hence by (2.5.9), $\delta_L^g(x, y)$ is a quasimetric as well. Thus we have obtained (EV2)$_M$. □

In short, in the above reasoning, we have two steps:

(1) The existence of weakly adapted metric d implies the hyperbolicity of g.
(2) The hyperbolicity of g implies the existence of an adapted metric ρ.

Notably that the original weakly adapted metric d may essentially differ from the adapted metric ρ. In fact, in Example 2.5.18, we are going to present an explicit example where no power of the original weakly adapted metric is bi-Lipschitz equivalent to any adapted metric.

Proof of Corollary 2.5.13 Note that $(T, \mathcal{B}) = (\tilde{T}^{hr,r}, \mathcal{B}_{\tilde{T}^{hr,r}})$ for any $r \in (0, 1)$. Assume that $(T, \mathcal{B})$ is hyperbolic. By Theorem 2.5.12, there exist $d \in \mathcal{D}(X, \mathcal{O})$, $M \ge 1$ and $\alpha > 0$ such that d is adapted to $(h_{1/2})^\alpha$. Since $(h_{1/2})^\alpha = h_{2-\alpha}$, we have the desired statement with $r = 2^{-\alpha}$. Conversely, with the existence of d, M and r, Theorem 2.5.12 implies that $(\tilde{T}^{hr,s}, \mathcal{B}_{\tilde{T}^{hr,s}})$ is hyperbolic for any $s \in (0, 1)$. Letting $s = r$, we see that $(T, \mathcal{B})$ is hyperbolic. □

To end this section, we are going to integrate the works by Elek[15] and Lau-Wang[25] into our framework.

Example 2.5.18 In Example 2.2.13, we have obtained a partition of a compact metric space in $\mathbb{R}^N$ corresponding to the hyperbolic graph constructed by Elek in [15]. In fact, we have obtained two graphs $(T, \tilde{\mathcal{B}})$ and $(T, \mathcal{B})$ satisfying $\tilde{\mathcal{B}} \supseteq \mathcal{B}$. The former coincides with Elek's graph and the latter is the resolution associated with the partition constructed from the Dyadic cubes. In this example, using Propositions 2.5.16 and 2.5.17, we are going to show the hyperbolicity of the graph $(T, \mathcal{B})$. The hyperbolicity of the original graph $(T, \tilde{\mathcal{B}})$ may be shown in a similar fashion.

Let X be a compact subset of $[0, 1]^N$ and let $(T, \mathcal{A}, \phi)$ be the tree associated with X constructed from the dyadic cubes in Example 2.2.13. Also let $K : T \to \mathcal{C}(X, \mathcal{O})$ be the partition of X parametrized by $(T, \mathcal{A}, \phi)$ given in Example 2.2.13. Set $g(w) = 2^{-m}$ if $w \in (T)_m$. Then $\Lambda_r^g = (T)_m$ if and only if $2^{-m} \le r < 2^{-m+1}$.

Let d_* be the Euclidean metric. Then for any $w \in \Lambda_r^g$,

$$\mathrm{diam}(K_w, d_*) \le \sqrt{N} r.$$

This shows (2.5.4). Moreover, if $w \in \Lambda_r^g$ and $K_w \cap B_{d_*}(x, cr) \ne \emptyset$, then $C(w) \subseteq B_{d_*}(x, (c + \sqrt{N})r)$. Note that $|C(w)|_N = 2^{-mN}$ for any $w \in (T)_m$, where $|\cdot|_N$ is the N-dimensional Lebesgue measure. Therefore, if $2^{-m} \le r < 2^{-m+1}$, then

$$\#(\{w | w \in \Lambda_r^g, B_{d_*}(x, cr) \cap K_w \ne \emptyset\}) \le \frac{|B_{d_*}(x, (c + \sqrt{N})r)|_N}{2^{-mN}}$$

$$= |B_{d_*}(0, 1)|_N (c + \sqrt{N})^N (2^m r)^N \le |B_{d_*}(0, 1)|_N (c + \sqrt{N})^N 2^N.$$

Therefore choosing $M \in \mathbb{N}$ so that $|B_{d_*}(0, 1)|_N (c + \sqrt{n})^n 2^n \le M + 1$, we have (2.5.5). Hence by Proposition 2.5.16, (the restriction of) d_* is weakly M-adapted to g. Since $(\widetilde{T}^{g, \frac{1}{2}}, \mathcal{B}_{\widetilde{T}^{g, \frac{1}{2}}}) = (T, \mathcal{B})$, Proposition 2.5.17 yields that $(T, \mathcal{B})$ is hyperbolic and its hyperbolic boundary coincides with X.

As we have mentioned above, in this example, the weakly adapted metric d_* is not necessarily adapted to any power of g. For example, let

$$X = [0, 1] \times \{0\} \bigcup \{(t, t) | t \in [0, 1]\} \bigcup$$

$$\left(\bigcup_{m \ge 1} \left\{ \left(\frac{1}{2^m}, s \right) \Big| s \in \left[0, \frac{1}{2^m} \right] \Big\backslash \left(\frac{1 - \epsilon_m}{2^{m+1}}, \frac{1 + \epsilon_m}{2^{m+1}} \right) \right\} \right),$$

where $\epsilon_m = \dfrac{1}{2^{m^2}}$. Set $x_m = \left(\dfrac{1}{2^m}, \dfrac{1 - \epsilon_m}{2^{m+1}} \right)$ and $y_m = \left(\dfrac{1}{2^m}, \dfrac{1 + \epsilon_m}{2^{m+1}} \right)$. Then $d_*(x_m, y_m) = \dfrac{\epsilon_m}{2^m}$ and $\delta_1^g(x_m, y_m) = \dfrac{1}{2^{m+1}}$. Then

$$\frac{d_*(x_m, y_m)}{\delta_1^{g^\alpha}(x_m, y_m)} = 2^{\alpha(m+1) - m^2 - m} \to 0$$

as $m \to \infty$ for any $\alpha > 0$. Thus for any $\alpha > 0$, the Euclidean metric is not bi-Lipschitz equivalent to any metric adapted to g^α.

Example 2.5.19 Let X be the self-similar set associated with the collection of contractions $\{F_1, \ldots, F_N\}$ and let $K : T^{(N)} \to X$ be the partition of K parametrized by $(T^{(N)}, \mathcal{A}^{(N)}, \phi)$ introduced in Example 2.2.7. We write $T = T^{(N)}$ for simplicity. In this example, we further assume that for any $i = 1, \ldots, N$, $F_i : \mathbb{R}^n \to \mathbb{R}^n$ is a similitude, i.e. there exist an orthogonal matrix A_i, $r_i \in (0, 1)$ and $a_i \in \mathbb{R}^n$ such that $F_i(x) = r_i A_i x + a_i$. Furthermore, we assume that the open set condition holds, i.e. there exists a nonempty open subset O of $\mathbb{R}^n$ such that $F_w(O) \subseteq O$ for any $w \in T$ and $F_w(O) \cap F_v(O) = \emptyset$ if $w, v \in T$ and $T_w \cap T_v = \emptyset$. Define $g(w) = r_{w_1} \cdots r_{w_m}$ for any $w = w_1 \ldots w_m \in T$. In this case, the conditions (2.5.4) and (2.5.5) have been known to hold for the Euclidean metric d_*. See [19, Proposition 1.5.8] for example. Hence Proposition 2.5.16 implies that d_* is weakly M-adapted to g for some $M \in \mathbb{N}$. Using Proposition 2.5.17, we see that $(\widetilde{T}^{g,r}, \mathcal{B}_{\widetilde{T}^{g,r}})$ is hyperbolic for any $r \in (0, 1)$ and hence the self-similar set X is the hyperbolic boundary of $(\widetilde{T}^{g,r}, \mathcal{B}_{\widetilde{T}^{g,r}})$. This fact has been shown by Lau and Wang in [25]. As in the previous example, the Euclidean metric is not necessarily a visual metric in this case.

Chapter 3
Relations of Weight Functions

3.1 Bi-Lipschitz Equivalence

In this section, we define the notion of bi-Lipschitz equivalence of weight functions. Originally the definition, Definition 3.1.1, only concerns the tree structure $(T, \mathcal{A}, \phi)$ and has nothing to do with a partition of a space. Under proper conditions, however, we will show that the bi-Lipschitz equivalence of weight functions is identified with

- absolutely continuity with uniformly bounded Radon-Nikodym derivative from below and above between measures in 3.1.1.
- usual bi-Lipschitz equivalence between metrics in 3.1.2.
- Ahlfors regularity of a measure with respect to a metric in 3.1.3.

As in the previous sections, $(T, \mathcal{A}, \phi)$ is a locally finite tree with a reference point ϕ, $(X, \mathcal{O})$ is a compact metrizable topological space with no isolated point and $K : T \to \mathcal{C}(X, \mathcal{O})$ is a partition of X parametrized by $(T, \mathcal{A}, \phi)$.

Definition 3.1.1 Two weight functions $g, h \in \mathcal{G}(T)$ are said to be bi-Lipschitz equivalent if and only if there exist $c_1, c_2 > 0$ such that

$$c_1 g(w) \leq h(w) \leq c_2 g(w)$$

for any $w \in T$. We write $g \underset{BL}{\sim} h$ if and only if g and h are bi-Lipschitz equivalent.

By the definition, we immediately have the next fact.

Proposition 3.1.2 *The relation* $\underset{BL}{\sim}$ *is an equivalent relation on* $\mathcal{G}(T)$.

© The Editor(s) (if applicable) and The Author(s), under exclusive license to Springer Nature Switzerland AG 2020
J. Kigami, *Geometry and Analysis of Metric Spaces via Weighted Partitions*, Lecture Notes in Mathematics 2265, https://doi.org/10.1007/978-3-030-54154-5_3

3.1.1   Bi-Lipschitz Equivalence of Measures

As we mentioned above, the bi-Lipschitz equivalence between weight functions can be identified with other properties depending on classes of weight functions of interest. First, we consider the case of weight functions associated with measures.

Definition 3.1.3 Let $\mu, \nu \in \mathcal{M}_P(X, \mathcal{O})$. We write $\mu \underset{AC}{\sim} \nu$ if and only if there exist $c_1, c_2 > 0$ such that

$$c_1 \mu(A) \le \nu(A) \le c_2 \mu(A) \tag{3.1.1}$$

for any Borel set $A \subseteq X$.

It is easy to see that $\underset{AC}{\sim}$ is an equivalence relation and $\mu \underset{AC}{\sim} \nu$ if and only if μ and ν are mutually absolutely continuous and the Radon-Nikodym derivative $\frac{d\nu}{d\mu}$ is uniformly bounded from below and above.

Theorem 3.1.4 *Assume that the partition $K : T \to \mathcal{C}(X, \mathcal{O})$ is strongly finite, i.e.*

$$\sup\{\#(\sigma^{-1}(x)) | x \in X\} < +\infty.$$

Let $\mu, \nu \in \mathcal{M}_P(X, \mathcal{O})$. Then $g_\mu \underset{BL}{\sim} g_\nu$ if and only if $\mu \underset{AC}{\sim} \nu$. Moreover, the natural map $\mathcal{M}_P(X, P)/\underset{AC}{\sim} \to \mathcal{G}(X)/\underset{BL}{\sim}$ given by $[g_\mu]_{\underset{BL}{\sim}}$ is injective, where $[\,\cdot\,]_{\underset{BL}{\sim}}$ is the equivalence class under $\underset{BL}{\sim}$.

Proof By (3.1.1), we see that $\alpha_1 \nu(K_w) \le \mu(K_w) \le \alpha_2 \nu(K_w)$ and hence $g_\mu \underset{BL}{\sim} g_\nu$. Conversely, if

$$c_1 \mu(K_w) \le \nu(K_w) \le c_2 \mu(K_w)$$

for any $w \in T$. Let $U \subset X$ be an open set. Assume that $U \ne X$. For any $x \in X$, there exists $w \in T$ such that $x \in K_w \subseteq U$. Moreover, if $K_w \subseteq U$, then there exists $m \in \{1, \ldots, |w|\}$ such that $K_{[w]_m} \subseteq U$ but $K_{[w]_{m-1}} \backslash U \ne \emptyset$. Therefore, if

$$T(U) = \{w | w \in T, K_w \subseteq U, K_{\pi(w)} \backslash U \ne \emptyset\},$$

then $T(U) \ne \emptyset$ and $U = \cup_{w \in T(U)} K_w$. Now, since K is strongly finite, there exists $N \in \mathbb{N}$ such that $\#(\sigma^{-1}(x)) \le N$ for any $x \in X$. Let $y \in U$. If $w(1), \ldots, w(m) \in T(U)$ are mutually different and $y \in K_{w(i)}$ for any $i = 1, \ldots, m$, then there exists $\omega(i) \in \Sigma_{w(i)}$ such that $\sigma(\omega(i)) = y$ for any $i = 1, \ldots, m$. If $i \ne j$, since $\Sigma_{w(i)} \cap \Sigma_{w(j)} = \emptyset$, it follows that $\omega(i) \ne \omega(j)$. Hence $m \le \#(\sigma^{-1}(y)) \le N$.

By Proposition A.1, we see that

$$\nu(U) \leq \sum_{w \in T(U)} \nu(K_w) \leq \sum_{w \in T(U)} c_2 \mu(K_w) \leq c_2 N \mu(U)$$

$$\mu(U) \leq \sum_{w \in T(U)} \mu(K_w) \leq \sum_{w \in T(U)} \frac{1}{c_1} \nu(K_w) \leq \frac{N}{c_1} \nu(U).$$

Hence letting $\alpha_1 = c_1/N$ and $\alpha_2 = c_2 N$, we have

$$\alpha_1 \mu(U) \leq \nu(U) \leq \alpha_2 \mu(U)$$

for any open set $U \subseteq X$. Since μ and ν are Radon measures, this yields (3.1.1). $\square$

3.1.2 Bi-Lipschitz Equivalence of Metrics

Under the tightness of weight functions defined below, we will translate bi-Lipschitz equivalence of weight functions to that of the visual pre-metrics associated with weight functions in Theorem 3.1.8. The tightness of a weight function ensures that δ_M^g is comparable with g, i.e the diameter with respect to δ_M^g of K_w is bi-Lipschitz equivalent to g.

Definition 3.1.5 A weight function g is called tight if and only if for any $M \geq 0$, there exists $c > 0$ such that

$$\sup_{x,y \in K_w} \delta_M^g(x, y) \geq cg(w)$$

for any $w \in T$.

Proposition 3.1.6 *Let g and h be weight functions. If g is tight and $g \underset{BL}{\sim} h$, then h is tight.*

Proof Since $g \underset{BL}{\sim} h$, there exist $\gamma_1, \gamma_2 > 0$ such that $\gamma_1 g(w) \leq h(w) \leq \gamma_2 g(w)$ for any $w \in T$. Therefore,

$$\gamma_1 D_M^g(x, y) \leq D_M^h(x, y) \leq \gamma_2 D_M^g(x, y)$$

for any $x, y \in X$ and $M \geq 0$. By Proposition 2.4.4, for any $M \geq 0$, there exist $c_1, c_2 > 0$ such that

$$c_1 \delta_M^g(x, y) \leq \delta_M^h(x, y) \leq c_2 \delta_M^g(x, y)$$

for any $x, y \in X$. Hence

$$\sup_{x,y \in K_w} \delta_M^h(x, y) \geq c_1 \sup_{x,y \in K_w} \delta_M^g(x, y) \geq c_1 c g(w) \geq c_1 c (\gamma_2)^{-1} h(w)$$

for any $w \in T$. Thus h is tight.   □

Any weight function induced by a metric is tight.

Proposition 3.1.7 *Let $d \in \mathcal{D}(X, \mathcal{O})$. Then g_d is tight.*

Proof Let $x, y \in X$ and let $(w(1), \ldots, w(M + 1)) \in \mathcal{CH}_K(x, y)$. Set $x_0 = x$ and $x_{M+1} = y$. For each $i = 1, \ldots, M$, choose $x_i \in K_{w(i)} \cap K_{w(i+1)}$. Then

$$\sum_{i=1}^{M+1} g_d(w(i)) \geq \sum_{i=1}^{M+1} d(x_{i-1}, x_i) \geq d(x, y).$$

Using this inequality and Proposition 2.4.4, we obtain

$$(M + 1)\delta_M^g(x, y) \geq D_M^{g_d}(x, y) \geq d(x, y)$$

and therefore $(M + 1) \sup_{x,y \in K_w} \delta_M^{g_d}(x, y) \geq g_d(w)$ for any $w \in T$. Thus g_d is tight.   □

Now we give geometric conditions which are equivalent to bi-Lipschitz equivalence of tight weight functions. The essential point is that bi-Lipschitz condition between weight functions is equivalent to that between the associated visual premetrics explicitly described in (BL2) and (BL3).

Theorem 3.1.8 *Let g and h be weight functions. Assume that both g and h are tight. Then the following conditions are equivalent:*

(BL) $g \underset{BL}{\sim} h.$

(BL1)   *There exist M_1, M_2 and $c > 0$ such that*

$$\delta_{M_1}^g(x, y) \leq c\delta_0^h(x, y) \quad and \quad \delta_{M_2}^h(x, y) \leq c\delta_0^g(x, y)$$

for any $x, y \in X$.

(BL2)   *There exist $c_1, c_2 > 0$ and $M \geq 0$ such that*

$$c_1 \delta_M^g(x, y) \leq \delta_M^h(x, y) \leq c_2 \delta_M^g(x, y)$$

for any $x, y \in X$.

(BL3)   *For any $M \geq 0$, there exist $c_1, c_2 > 0$ such that*

$$c_1 \delta_M^g(x, y) \leq \delta_M^h(x, y) \leq c_2 \delta_M^g(x, y)$$

for any $x, y \in X$.

Before a proof of this theorem, we state a notable corollary, which shows that when weight functions are induced by adapted metrics, bi-Lipschitz equivalence of weight functions exactly corresponds to the usual bi-Lipschitz equivalence of metrics.

Definition 3.1.9

(1) Let $d, \rho \in \mathcal{D}(X, \mathcal{O})$. d and ρ are said to be bi-Lipschitz equivalent, $d \underset{BL}{\sim} \rho$ for short, if and only if there exist $c_1, c_2 > 0$ such that

$$c_1 d(x, y) \le \rho(x, y) \le c_2 d(x, y)$$

for any $x, y \in X$.

(2) Define

$$\mathcal{D}_A(X, \mathcal{O}) = \{d \mid d \in \mathcal{D}(X, \mathcal{O}), d \text{ is adapted.}\}$$

Corollary 3.1.10 *Let $d, \rho \in \mathcal{D}_A(X, \mathcal{O})$. Then $g_d \underset{BL}{\sim} g_\rho$ if and only if $d \underset{BL}{\sim} \rho$. In particular, the correspondence of $[d]_{\underset{BL}{\sim}}$ with $[g_d]_{\underset{BL}{\sim}}$ gives an well-defined injective map $\mathcal{D}_A(X, \mathcal{O})/_{\underset{BL}{\sim}} \to \mathcal{G}(X)/_{\underset{BL}{\sim}}$.*

Now we start to prove Theorem 3.1.8 and its corollary.

Lemma 3.1.11 *Let h be a weight function. If $x \in K_w$ and $K_w \backslash U_0^h(x, s) \ne \emptyset$, then $s < h(w)$.*

Proof Since $x \in K_w$, there exists $\omega \in \Sigma$ such that $\sigma(\omega) = x$ and $[\omega]_m = w$, where $m = |w|$. Then there exists a unique k such that $[\omega]_k \in \Lambda_s^h$. Since $x \in K_{[\omega]_k}$, it follows that $K_{[\omega]_k} \subseteq U_0^h(x, s)$. On the other hand, if $n \le m$, we have $K_{[\omega]_n} \supseteq K_w$ and hence $K_{[\omega]_n} \backslash U_0^h(x, s) \ne \emptyset$. This shows that $k < m$. Consequently $h(w) \ge h(\pi([\omega]_k)) > s$. □

Proposition 3.1.12 *Let g and h be weight functions. Assume that g is tight. Let $M \ge 0$. If there exists $\alpha > 0$ such that*

$$\alpha \delta_M^g(x, y) \le \delta_0^h(x, y) \tag{3.1.2}$$

for any $x, y \in X$. Then there exists $c > 0$ such that

$$g(w) \le ch(w)$$

for any $w \in T$.

Proof Since g is tight, there exists $c_1 > 0$ such that $\sup_{x,y \in K_w} \delta_M^g(x, y) \ge c_1 g(w)$. Assume that $K_w \subseteq U_M^g(x, \beta g(w))$ for any $x \in K_w$. Then for any $x, y \in K_w$,

$\delta_M^g(x, y) \le \beta g(w)$ and hence

$$\beta g(w) \ge \sup_{x,y} \delta_M^g(x, y) \ge c_1 g(w).$$

This implies $\beta \ge c_1$. So, letting $\beta < c_1$, we see that, for any $w \in T$,

$$K_w \setminus U_M^g(x, \beta g(w)) \ne \emptyset$$

for some $x \in K_w$. On the other hand, by (3.1.2), $U_M^g(x, s) \supseteq U_0^h(x, \alpha s)$ for any $x \in X$ and $s \ge 0$. Therefore,

$$K_w \setminus U_0^h(x, \alpha\beta g(w)) \ne \emptyset.$$

By Lemma 3.1.11, we have $\alpha\beta g(w) \le h(w)$.   □

Remark By the above discussion, we see that g is tight if and only if for any $M \ge 0$, there exists $\beta > 0$ such that for any $w \in T$,

$$K_w \setminus U_M^g(x, \beta g(w)) \ne \emptyset$$

for some $x \in K_w$.

Lemma 3.1.13 *Let g and h be weight functions. Assume that g is tight. Then the following conditions are equivalent:*

(A) *There exists $c > 0$ such that $g(w) \le ch(w)$ for any $w \in T$.*
(B) *For any $M, N \ge 0$ with $N \ge M$, there exists $c > 0$ such that*

$$\delta_N^g(x, y) \le c\delta_M^h(x, y)$$

for any $x, y \in X$.
(C) *There exist $M, N \ge 0$ and $c > 0$ such that $N \ge M$ and*

$$\delta_N^g(x, y) \le c\delta_M^h(x, y)$$

for any $x, y \in X$.

Proof (A) implies

$$D_M^g(x, y) \le cD_M^h(x, y) \tag{3.1.3}$$

for any $x, y \in X$ and $M \ge 0$. By Proposition 2.4.4, we see

$$\delta_M^g(x, y) \le c(M + 1)\delta_M^h(x, y)$$

for any $x, y \in X$. Since $\delta_N^g(x, y) \le \delta_M^g(x, y)$ if $N \ge M$, we have (B). Obviously (B) implies (C). Now assume (C). Then we have $\delta_N^g(x, y) \le c\delta_0^h(x, y)$. Hence Proposition 3.1.12 yields (A). $\qquad\square$

Proof of Theorem 3.1.8 Lemma 3.1.13 immediately implies the desired statement. $\qquad\square$

Proof of Corollary 3.1.10 Since d and ρ are adapted, by (2.4.1), there exist $M \ge 1$ and $c > 0$ such that

$$c\delta_M^d(x, y) \le d(x, y) \le \delta_M^d(x, y), \tag{3.1.4}$$

$$c\delta_M^\rho(x, y) \le \rho(x, y) \le \delta_M^\rho(x, y) \tag{3.1.5}$$

for any $x, y \in X$. Assume $g_d \underset{BL}{\sim} g_\rho$. Since g_d and g_ρ are tight, we have (BL3) by Theorem 3.1.8. Hence by (3.1.4) and (3.1.5), $d(\cdot, \cdot)$ and $\rho(\cdot, \cdot)$ are bi-Lipschitz equivalent as metrics. The converse direction is immediate. $\qquad\square$

Roughly, the next theorem states that an adapted metric d is adapted to a weight function g if and only $g_d \underset{BL}{\sim} g$.

Theorem 3.1.14 *Let $d \in \mathcal{D}(X, \mathcal{O})$ and let g be a weight function. Then d is adapted to g and g is tight if and only if $g_d \underset{BL}{\sim} g$ and $d \in \mathcal{D}_A(X, \mathcal{O})$.*

Proof If d is M-adapted to g for some $M \ge 1$, then by (ADa), there exists $c > 0$ such that $g_d(w) \le cg(w)$ for any $w \in K_w$. Moreover, (2.4.1) implies

$$d(x, y) \ge c_2 \delta_M^g(x, y)$$

for any $x, y \in X$, where c_2 is independent of x and y. Hence the tightness of g shows that there exists $c' > 0$ such that

$$g_d(w) \ge c_2 \sup_{x, y \in K_w} \delta_M^g(x, y) \ge c'g(w).$$

Thus $g_d \underset{BL}{\sim} g$. Moreover, by Proposition 2.4.8, d is adapted. Conversely, assume that d is adapted and $g_d \underset{BL}{\sim} g$. Then there exist $c_1, c_2 > 0$ such that $c_1 g_d(w) \le g(w) \le c_2 g_d(w)$ for any $w \in T$. This implies, for any $M \ge 0$,

$$c_1 D_M^{g_d}(x, y) \le D_M^g(x, y) \le c_2 D_M^{g_d}(x, y)$$

for any $x, y \in X$. By Proposition 2.4.4,

$$c_1(M+1)^{-1}\delta_M^{g_d}(x, y) \le \delta_M^g(x, y) \le c_2(M+1)\delta_M^{g_d}(x, y) \tag{3.1.6}$$

for any $x, y \in X$. Since g_d is tight, there exists $c > 0$ such that

$$\sup_{x,y \in K_w} \delta_M^g(x, y) \geq c_1 (M + 1)^{-1} \sup_{x,y \in K_w} \delta_M^{g_d}(x, y) \geq c g_d(w) \geq c(c_2)^{-1} g(w)$$

for any $w \in K_w$. Thus g is tight. If d is M_*-adapted to g_d, then (3.1.6) shows that d is M_*-adapted to g as well.   □

3.1.3   Bi-Lipschitz Equivalence Between Measures and Metrics

Finally in this sub-section, we consider what happens if the weight function associated with a measure is bi-Lipschitz equivalent to the weight function associated with a metric.

To state our theorem, we need the following notions.

Definition 3.1.15

(1) A weight function $g : T \to (0, 1]$ is said to be uniformly finite if $\sup\{\#(\Lambda_{s,1}^g(w)) | s \in (0, 1], w \in \Lambda_s^g\} < +\infty$.
(2) A function $f : T \to (0, \infty)$ is called sub-exponential if and only if there exist $m \geq 0$ and $c_1 \in (0, 1)$ such that $f(v) \leq c_1 f(w)$ for any $w \in T$ and $v \in T_w$ with $|v| \geq |w| + m$. f is called super-exponential if and only if there exists $c_2 \in (0, 1)$ such that $f(v) \geq c_2 f(w)$ for any $w \in T$ and $v \in S(w)$. f is called exponential if it is both sub-exponential and super-exponential.

The following proposition and the lemma are immediate consequences of the above definitions.

Proposition 3.1.16 *Let h be a weight function. Then h is super-exponential if and only if there exists $c \geq 1$ such that $ch(w) \geq s \geq h(w)$ whenever $w \in \Lambda_s^h$.*

Proof Assume that h is super-exponential. Then there exists $c_2 \in (0, 1)$ such that $h(w) \geq c_2 h(\pi(w))$ for any $w \in T$. If $w \in \Lambda_s^h$, then $h(\pi(w)) > s \geq h(w)$. This implies $(c_2)^{-1} h(w) \geq s \geq h(w)$.

Conversely, assume that $ch(w) \geq s \geq h(w)$ for any w and s with $w \in \Lambda_s^h$. If $h(\pi(w)) > ch(w)$, then $w \in \Lambda_t^h$ for any $t \in (ch(w), h(\pi(w)))$. This contradicts the assumption that $ch(w) \geq t \geq h(w)$. Hence $h(\pi(w)) \leq ch(w)$ for any $w \in T$. Thus h is super-exponential.   □

Lemma 3.1.17 *If a weight function $g : T \to (0, 1]$ is uniformly finite, then*

$$\sup\{\#(\Lambda_{s,M}(x)) | x \in X, s \in (0, 1]\} < +\infty$$

for any $M \geq 0$.

Proof Let $C = \sup\{\#(\Lambda_{s,1}(w)) | s \in (0, 1], w \in \Lambda_s\}$. Then $\#(\Lambda_{s,M}(x)) \leq C + C^2 + \ldots + C^{M+1}$.   □

Definition 3.1.18 Let $\alpha > 0$. A Radon measure μ on X is said to be α-Ahlfors regular with respect to $d \in \mathcal{D}(X, \mathcal{O})$ if and only if there exist $C_1, C_2 > 0$ such that

$$C_1 r^\alpha \leq \mu(B_d(x, r)) \leq C_2 r^\alpha \tag{3.1.7}$$

for any $r \in [0, \text{diam}(X, d)]$.

Definition 3.1.19 Let $g : T \to (0, 1]$ be a weight function. We say that K has thick interior with respect to g, or g is thick for short, if and only if there exist $M \geq 1$ and $\alpha > 0$ such that $K_w \supseteq U_M^g(x, \alpha s)$ for some $x \in K_w$ if $s \in (0, 1]$ and $w \in \Lambda_s^g$.

The value of the integer $M \geq 1$ is not crucial in the above definition. In Proposition 3.2.1, we will show if the condition of the above definition holds for a particular $M \geq 1$, then it holds for all $M \geq 1$.

The thickness is invariant under the bi-Lipschitz equivalence of weight functions as follows.

Proposition 3.1.20 *Let g and h be weight functions. If g is thick and $g \underset{BL}{\sim} h$, then h is thick.*

Since we need further results on the thickness of weight functions, we postpone a proof of this proposition until the next section.

Now we give the main theorem of this sub-section.

Theorem 3.1.21 *Let $d \in \mathcal{D}_A(X, \mathcal{O})$ and let $\mu \in \mathcal{M}_P(X, \mathcal{O})$. Assume that K is minimal and g_d is super-exponential and thick. Then $(g_d)^\alpha \underset{BL}{\sim} g_\mu$ and g_d is uniformly finite if and only if μ is α-Ahlfors regular with respect to d. Moreover, if either/both of the these two conditions is/are satisfied and $\#(S(w)) \geq 2$ for any $w \in T$, then g_μ and g_d are exponential.*

By the same reason as Proposition 3.1.20, a proof of this theorem will be given at the end of Sect. 3.3.

3.2 Thickness of Weight Functions

In this section, we study conditions for a weight function being thick and the relation between the notions "thick" and "tight". For instance in Theorem 3.2.3 we present topological condition (TH1) ensuring that all super-exponential weight functions are thick. In particular, this is the case for partitions of S^2 discussed in Sect. 1.2 because partitions satisfying (1.2.2) are minimal and the condition (TH) in Sect. 1.2 yields the condition (TH1). Moreover in this case, all super-exponential weight functions are tight as well by Corollary 3.2.5.

Proposition 3.2.1 *g is thick if and only if for any $M \geq 0$, there exists $\beta > 0$ such that, for any $w \in T$, $K_w \supseteq U_M^g(x, \beta g(\pi(w)))$ for some $x \in K_w$.*

Proof Assume that g is thick. By induction, we are going to show the following claim $(C)_M$ holds for any $M \geq 1$:

$(C)_M$ There exists $\alpha_M > 0$ such that, for any $s \in (0, 1]$ and $w \in \Lambda_s^g$, one find $x \in K_w$ satisfying $K_w \supseteq U_M(x, \alpha_M s)$.

Proof of $(C)_M$ Since g is thick, $(C)_M$ holds for some $M \geq 1$. Since $U_1^g(x, s) \subseteq U_M^g(x, s)$ if $M \geq 1$, $(C)_1$ holds as well. Now, suppose that $(C)_M$ holds. Let $w \in \Lambda_s^g$ and choose x as in $(C)_M$. Then there exists $v \in \Lambda_{\alpha_M s}^g$ such that $v \in T_w$ and $x \in K_v$. Applying $(C)_M$ again, we find $y \in K_v$ such that $K_v \supseteq U_M^g(y, (\alpha_M)^2 s)$. Since $M \geq 1$, it follows that $U_{M+1}^g(y, (\alpha_M)^2 s) \subseteq U_M^g(x, \alpha_M s) \subseteq K_w$. Therefore, letting $\alpha_{M+1} = (\alpha_M)^2$, we have obtained $(C)_{M+1}$. Thus we have shown $(C)_M$ for any $M \geq 1$.   □

Next, fix $M \geq 1$ and write $\alpha = \alpha_M$. Note that $u \in \Lambda_s^g$ if and only if $g(u) \leq s < g(\pi(u))$. Let $w \in T$. Fix $\epsilon \in (0, 1)$. Assume that $g(\pi(w)) > g(w)$. There exists s_* such that $g(w) \leq s_* < g(\pi(w))$ and $s_* > \epsilon g(\pi(w))$. Hence we obtain

$$K_w \supseteq U_M^g(x, \alpha s_*) \supseteq U_M^g(x, \alpha \epsilon g(\pi(w))).$$

If $g(w) = g(\pi(w))$, then there exists $v \in T_w$ such that $g(v) < g(\pi(v)) = g(w) = g(\pi(w))$. Choosing s_* so that $g(v) \leq s_* < g(\pi(v)) = g(\pi(w))$ and $\epsilon g(\pi(w)) < s_*$, we obtain

$$K_w \supseteq K_v \supseteq U_M^g(x, \alpha s_*) \supseteq U_M^g(x, \alpha \epsilon g(\pi(w))).$$

Letting $\beta = \alpha \epsilon$, we obtain the desired statement.

Conversely, assume that for any $M \geq 0$, there exists $\beta > 0$ such that, for any $w \in T$, $K_w \supseteq U_M^g(x, \beta g(\pi(w)))$ for some $x \in K_w$. If $w \in \Lambda_s$, then $g(w) \leq s < g(\pi(w))$. Therefore $K_w \supseteq U_M^g(x, \beta s)$. This implies that g is thick.   □

Proposition 3.2.2 *Assume that K is minimal. Let $g : T \to (0, 1]$ be a weight function. Then g is thick if and only if, for any $M \geq 0$, there exists $\gamma > 0$ such that, for any $w \in T$, $O_w \supseteq U_M^g(x, \gamma g(\pi(w)))$ for some $x \in O_w$.*

Proof Assume that g is thick. By Proposition 3.2.1, for any $M \geq 0$, we may choose $\alpha > 0$ so that for any $w \in T$, there exists $x \in K_w$ such that $K_w \supseteq U_{M+1}^g(x, \alpha g(\pi(w)))$. Set $s_w = g(\pi(w))$. Let $y \in U_M^g(x, \alpha s_w) \setminus O_w$. There exists $v \in (T)_{|w|}$ such that $y \in K_v$ and $w \neq v$. Then we find $v_* \in T_v \cap \Lambda_{\alpha s_w}^g$ satisfying $y \in K_{v_*}$. Since $K_{v_*} \cap U_M^g(x, \alpha s_w) \neq \emptyset$, we have

$$K_{v_*} \subseteq U_{M+1}^g(x, \alpha s_w) \subseteq K_w.$$

Therefore, $K_{v_*} \subseteq \cup_{w' \in T_w, |w'| = |v_*|} K_{w'}$. This implies that $O_{v_*} = \emptyset$, which contradicts the fact that K is minimal. So, $U_M^g(x, \alpha s_w) \setminus O_w = \emptyset$ and hence $U_M^g(x, \alpha s_w) \subseteq O_w$.

The converse direction is immediate.   □

Using Proposition 3.2.1, we give a proof of Proposition 3.1.20.

Proof of Proposition 3.1.20 By Proposition 3.2.1, there exists $\beta > 0$ such that for any $w \in T$, $K_w \supseteq U_M^g(x, \beta g(\pi(w)))$ for some $x \in K_w$. On the other hand, since there exist $c_1, c_2 > 0$ such that $c_1 h(w) \leq g(w) \leq c_2 h(w)$ for any $w \in T$, it follows that $D_M^g(x, y) \leq c_2 D_M^h(x, y)$ for any $x, y \in X$. Proposition 2.4.4 implies that there exists $\alpha > 0$ such that $\alpha \delta_M^g(x, y) \leq \delta_M^h(x, y)$ for any $x, y \in X$. Hence $U_M^h(x, \alpha s) \subseteq U_M^g(x, s)$ for any $x \in X$ and $s \in (0, 1]$. Combining them, we see that

$$K_w \supseteq U_M^g(x, \beta g(\pi(w))) \supseteq U_M^h(x, \alpha \beta g(\pi(w))) \supseteq U_M^h(x, \alpha \beta c_1 h(\pi(w))).$$

Thus by Proposition 3.2.1, h is thick. □

Theorem 3.2.3 *Assume that K is minimal. Define $h_* : T \to (0, 1]$ by $h_*(w) = 2^{-|w|}$ for any $w \in T$. Then the following conditions are equivalent:*

(TH1)

$$\sup_{w \in T} \min \left\{ |v_*| - |w| \big| v_* \in T_w, K_{v_*} \subseteq O_w \right\} < \infty.$$

(TH2) *Every super-exponential weight function is thick.*

(TH3) *There exists a thick sub-exponential weight function.*

(TH4) *The weight function h_* is thick.*

Proof (TH1) $\Rightarrow$ (TH2): Assume (TH1). Let m be the supremum in (TH1). Let g be a super-exponential weight function. Then there exists $\lambda \in (0, 1)$ such that $g(w) \geq \lambda g(\pi(w))$ for any $w \in T$. Let $w \in \Lambda_s^g$. By (TH1), there exists $v_* \in T_w \cap (T)_{|w|+m}$ such that $K_{v_*} \subseteq O_w$. For any $v \in T_w \cap (T)_{|w|+m}$,

$$g(v) \geq \lambda^m g(w) \geq \lambda^{m+1} g(\pi(w)) > \lambda^{m+1} s.$$

Choose $x \in O_{v_*}$. Let $u \in \Lambda_{\lambda^{m+1}s,1}(x)$. Then there exists $v' \in \Lambda_{\lambda^{m+1}s,0}(x)$ such that $K_{v'} \cap K_u \neq \emptyset$. Since $g(v_*) > \lambda^{m+1}s$ and $x \in O_{v_*}$, it follows that $v' \in T_{v_*}$. Therefore $K_u \cap O_w \supseteq K_u \cap K_{v_*} \neq \emptyset$. This implies that either $u \in T_w$ or $w \in T_u$. Since $g(w) > \lambda^{m+1}s$, it follows that $u \in T_w$. Thus we have shown that $\Lambda_{\lambda^{m+1}s,1}(x) \subseteq T_w$. Hence

$$U_1^g(x, \lambda^{m+1}s) \subseteq K_w.$$

This shows that g is thick.

(TH2) $\Rightarrow$ (TH4): Apparently h_* is an exponential weight function. Hence by (TH2), it is thick.

(TH4) $\Rightarrow$ (TH3): Since h_* is exponential and thick, we have (TH3).

(TH3) $\Rightarrow$ (TH1): Assume that g is a thick sub-exponential weight function. Proposition 3.2.2 shows that there exist $\gamma \in (0, 1)$ and $M \geq 1$ such that for any $w \in T$, $O_w \supseteq U_M^g(x, \gamma g(\pi(w)))$. Choose $v_* \in \Lambda_{\gamma g(\pi(w)),0}^g(x)$. Then $K_{v_*} \subseteq U_M^g(x, \gamma g(\pi(w))) \subseteq O_w$ and $g(\pi(v_*)) > \gamma g(\pi(w)) \geq \gamma g(w)$. Since g

is sub-exponential, there exist $k \geq 1$ and $\eta \in (0, 1)$ such that $g(u) \leq \eta g(v)$ if $v \in T_v$ and $|u| \geq |w| + k$. Choose l so that $\eta^l < \gamma$ and set $m = kl + 1$. Since $g(\pi(v_*)) > \eta^l g(w)$, we see that $|\pi(v_*)| \leq |w| + m - 1$. Therefore, $|v_*| \leq |w| + m$ and hence we have (TH1).   □

Theorem 3.2.4 *Let $g : T \to (0, 1]$ be a weight function. Assume that $K : T \to \mathcal{C}(X, \mathcal{O})$ is minimal, that there exists $\lambda \in (0, 1)$ such that if $B_w = \emptyset$, then $\#(T_w \cap \Lambda^g_{\lambda g(w)}) \geq 2$ and that g is thick. Then g is tight.*

Proof By Proposition 3.2.2, there exists γ such that, for any $v \in T$, $O_v \supseteq U^g_M(x, \gamma g(\pi(v)))$ for some $x \in K_v$. First suppose that $B_w \neq \emptyset$. Then there exists $x \in K_w$ such that $O_w \supseteq U^g_M(x, \gamma g(\pi(w)))$. For any $y \in B_w$, it follows that $\delta^g_M(x, y) > \gamma g(\pi(w))$. Thus

$$\sup_{x,y \in K_w} \delta^g_M(x, y) \geq \gamma g(\pi(w)) \geq \gamma g(w).$$

Next if $B_w = \emptyset$, then there exists $u \neq v \in T_w \cap \Lambda^g_{\lambda g(w)}$. If $B_u \neq \emptyset$, then the above discussion implies

$$\sup_{x,y \in K_w} \delta^g_M(x, y) \geq \sup_{x,y \in K_v} \delta^g_M(x, y) \geq \gamma g(\pi(v)) \geq \gamma \lambda g(w).$$

If $B_u = \emptyset$, then $\delta^g_M(x, y) \geq \lambda g(w)$ for any $(x, y) \in K_u \times K_v$. Thus for any $w \in T$, we conclude that

$$\sup_{x,y \in K_w} \delta^g_M(x, y) \geq \gamma \lambda g(w).$$

□

The above theorem immediately implies the following corollary.

Corollary 3.2.5 *Assume that $(X, \mathcal{O})$ is connected and K is minimal. If g is thick, then g is tight.*

3.3 Volume Doubling Property

In this section, we introduce the notion of a relation called "gentle" written as $\underset{\text{GE}}{\sim}$ between weight functions. This relation is not an equivalence relation in general. In Sect. 3.5, however, it will be shown to be an equivalence relation among exponential weight functions. As was the case of the bi-Lipschitz equivalence, the gentleness will be identified with other properties depending on the classes of weight functions of interest. In particular, we are going to show that the volume doubling property of a measure with respect to a metric is equivalent to the gentleness of the associated weight functions.

As in the previous sections, $(T, \mathcal{A}, \phi)$ is a locally finite tree with a reference point ϕ, $(X, \mathcal{O})$ is a compact metrizable topological space with no isolated point and $K : T \to \mathcal{C}(X, \mathcal{O})$ is a partition of X parametrized by $(T, \mathcal{A}, \phi)$.

The notion of the gentleness of a weight function to another weight function is defined as follows.

Remark In the case of the natural partition of a self-similar set in Example 2.2.7, the main results of this section, Theorems 3.3.7 and 3.3.9 have been obtained in [20].

Definition 3.3.1 Let $g : T \to (0, 1]$ be a weight function. A function $f : T \to (0, \infty)$ is said to be gentle with respect to g if and only if there exists $c_G > 0$ such that $f(v) \le c_G f(w)$ whenever $w, v \in \Lambda_s^g$ and $K_w \cap K_v \ne \emptyset$ for some $s \in (0, 1]$. We write $f \underset{GE}{\sim} g$ if and only if f is gentle with respect to g.

Alternatively, we have a simpler version of the definition of gentleness under a mild restriction.

Proposition 3.3.2 *Let $g : T \to (0, 1]$ be an exponential weight function. Let $f : T \to (0, \infty)$. Assume that $f(w) \le f(\pi(w))$ for any $w \in T$ and f is super-exponential. Then f is gentle with respect to g if and only if there exists $c > 0$ such that $f(v) \le cf(w)$ whenever $g(v) \le g(w)$ and $K_v \cap K_w \ne \emptyset$.*

Proof By the assumption, there exist $c_1, c_2 > 0$ and $m \ge 1$ such that $f(v) \ge c_2 f(w)$, $g(v) \ge c_2 g(w)$ and $g(u) \le c_1 g(w)$ for any $w \in T$, $v \in S(w)$ and $u \in T_w$ with $|u| \ge |w| + m$.

First suppose that f is gentle with respect to g. Then there exists $c > 0$ such that $f(v') \le cf(w')$ whenever $w', v' \in \Lambda_s^g$ and $K_{w'} \cap K_{v'} \ne \emptyset$ for some $s \in (0, 1]$. Assume that $g(v) \le g(w)$ and $K_v \cap K_w \ne \emptyset$. There exists $u \in T_w$ such that $K_u \cap K_v \ne \emptyset$ and $g(\pi(u)) > g(v) \ge g(u)$. Moreover, $g(\pi([v]_m)) > g([v]_m) = g(v)$ for some $m \in [0, |v|]$. Then $[v]_m, u \in \Lambda_{g(v)}^g$ and hence $f(v) \le f([v]_m) \le cf(u) \le cf(w)$.

Conversely, assume that $f(v') \le cf(w')$ whenever $g(v') \le g(w')$ and $K_{v'} \cap K_{w'} \ne \emptyset$. Let $w, v \in \Lambda_s^g$ with $K_w \cap K_v \ne \emptyset$. If $g(v) \le g(w)$, then $f(v) \le cf(w)$. Suppose that $g(v) > g(w)$. Since g is super-exponential, we see that

$$s \ge g(w) \ge c_2 g(\pi(w)) \ge c_2 s \ge c_2 g(v).$$

Set $N = \min\{n | c_2 \ge c_1^n\}$. Choose $u \in T_v$ so that $K_u \cap K_w \ne \emptyset$ and $|u| = |v| + Nm$. Then $g(w) \ge c_2 g(v) \ge (c_1)^N g(v) \ge g(u)$. This implies $f(u) \le cf(w)$. Since $f(u) \ge (c_2)^{Nm} f(v)$, we have $f(v) \le c(c_2)^{-Nm} f(w)$. Therefore, f is gentle with respect to g. $\square$

The following is the standard version of the definition of the volume doubling property.

Definition 3.3.3 Let μ be a Radon measure on $(X, \mathcal{O})$ and let $d \in \mathcal{D}(X, O)$. μ is said to have the volume doubling property with respect to the metric d if and only if there exists $C > 0$ such that

$$\mu(B_d(x, 2r)) \leq C\mu(B_d(x, r))$$

for any $x \in X$ and any $r > 0$.

Since $(X, \mathcal{O})$ has no isolated point, if a Radon measure μ has the volume doubling property with respect to some $d \in \mathcal{D}(X, \mathcal{O})$, then the normalized version of μ, $\mu/\mu(X)$, belongs to $\mathcal{M}_P(X, O)$. Taking this fact into account, we are mainly interested in (the normalized version of) a Radon measure in $\mathcal{M}_P(X, \mathcal{O})$.

The main theorem of this section is as follows.

Theorem 3.3.4 *Let $d \in \mathcal{D}(X, \mathcal{O})$ and let $\mu \in \mathcal{M}_P(X, \mathcal{O})$. Assume that d is adapted, that g_d is thick, exponential and uniformly finite and that μ is exponential. Then μ has the volume doubling property with respect to d if and only if $g_d \underset{\text{GE}}{\sim} g_\mu$.*

This theorem says that the volume doubling property equals the gentleness in the world of weight functions having certain regularities. We will verify this theorem as an immediate corollary of Theorems 3.3.7 and 3.3.9.

To describe a refined version of Theorem 3.3.4, we define the notion of volume doubling property of a measure with respect to a weight function g as well by means of "balls" $U_M^g(x, s)$.

Definition 3.3.5 Let $\mu \in \mathcal{M}_P(X, \mathcal{O})$ and let g be a weight function. For $M \geq 1$, we say μ has M-volume doubling property with respect to g if and only if there exist $\gamma \in (0, 1)$ and $\beta \geq 1$ such that $\mu(U_M^g(x, s)) \leq \beta\mu(U_M^g(x, \gamma s))$ for any $x \in X$ and any $s \in (0, 1]$.

It is rather annoying that the notion of "volume doubling property" of a measures with respect to a weight function depends on the value $M \geq 1$. Under certain conditions including exponentiality and the thickness, however, we will show that if μ has M-volume doubling property for some $M \geq 1$, then it has M-volume doubling property for all $M \geq 1$ in Theorem 3.3.9.

Naturally, if a metric is adapted to a weight function, the volume doubling with respect to the metric and that with respect to the weight function are virtually the same as is seen in the next proposition.

Proposition 3.3.6 *Let $d \in \mathcal{D}(X, \mathcal{O})$, let $\mu \in \mathcal{M}_P(X, \mathcal{O})$ and let g be a weight function. Assume that d is adapted to g. Then μ has the volume doubling property with respect to d if and only if there exists $M_* \geq 1$ such that μ has M-volume doubling property with respect to g for any $M \geq M_*$.*

Proof Since d is adapted to g, for sufficiently large M, there exist $\alpha_1, \alpha_2 > 0$ such that

$$U_M^g(x, \alpha_1 s) \subseteq B_d(x, s) \subseteq U_M^g(x, \alpha_2 s)$$

for any $x \in X$ and $s \in (0, 1]$. Suppose that μ has the volume doubling property with respect to d. Then there exists $\lambda > 1$ such that

$$\mu(B_d(x, 2^m r)) \leq \lambda^m \mu(B_d(x, r))$$

for any $x \in X$ and $r \geq 0$. Hence

$$\mu(U_M^g(x, \alpha_1 2^m r)) \subseteq \lambda^m \mu(U_M^g(x, \alpha_2 r)).$$

Choosing m so that $\alpha_1 2^m > \alpha_2$, we see that μ has M-volume doubling property with respect to g if M is sufficiently large. Converse direction is more or less similar. $\square$

By the above proposition, as far as we confine ourselves to adapted metrics, it is enough to consider the volume doubling property of a measure with respect to a weight function. Thus we are going to investigate relations between "the volume doubling property with respect to a weight function" and other conditions like a weight function being exponential, a weight function being uniformly finite, a measure being super-exponential, and a measure being gentle with respect to a weight function. To begin with, we show that the last four conditions imply the volume doubling property of μ with respect to g.

Theorem 3.3.7 *Let $g : T \to (0, 1]$ be a weight function and let $\mu \in \mathcal{M}_P(X, \mathcal{O})$. Assume that g is exponential, that g is uniformly finite, that μ is gentle with respect to g and that μ is super-exponential. Then μ has M-volume doubling property with respect to g for any $M \geq 1$.*

Hereafter in this section, we are going to omit g in notations if no confusion may occur. For example, we write Λ_s, $\Lambda_{s,M}(w)$, $\Lambda_{s,M}(w)$ and $U_M(x, s)$ in place of Λ_s^g, $\Lambda_{s,M}^g(w)$, $\Lambda_{s,M}^g(x)$ and $U_M^g(x, s)$ respectively.

The following lemma is a step to prove the above theorem.

Lemma 3.3.8 *Let $g : T \to (0, 1]$ be a weight function and let $\mu \in \mathcal{M}_P(X, \mathcal{O})$. For $s \in (0, 1]$, $\lambda > 1$, $k \geq 1$ and $c > 0$, define*

$$\Theta(s, \lambda, k, c) = \{v | v \in \Lambda_s, \mu(K_u) \leq c\mu(K_v) \text{ for any } u \in \Lambda_{\lambda s, k}((v)_{\lambda s})\},$$

where $(v)_{\lambda s}$ is the unique element of $\{[v]_n | 0 \leq n \leq |v|\} \cap \Lambda_{\lambda s}$. Assume that g is uniformly finite and that there exist $N \geq 1$, $\lambda > 1$ and $c > 0$ such that $\Lambda_{s,N}(w) \cap \Theta(s, \lambda, 2N + 1, c) \neq \emptyset$ for any $s \in (0, 1]$ and $w \in \Lambda_s$. Then μ has the N-volume doubling property with respect to g.

Proof Let $w \in \Lambda_{s,0}(x)$ and let $v \in \Lambda_{s,N}(w) \cap \Theta(s, \lambda, 2N + 1, c)$. If $u \in \Lambda_{\lambda s, N}(x)$, then $u \in \Lambda_{\lambda s, 2N+1}((v)_{\lambda s})$. Moreover, since $v \in \Lambda_{s,N}(x)$, we see that

$$\mu(K_u) \leq c\mu(K_v) \leq c\mu(U_N(x, s)).$$

Therefore,

$$\mu(U_N(x, \lambda s)) \leq \sum_{u \in \Lambda_{\lambda s, N}(x)} \mu(K_u) \leq \#(\Lambda_{\lambda s, N}(x)) c \mu(U_N(x, s)).$$

Since g is uniformly finite, Lemma 3.1.17 shows that $\#(\Lambda_{\lambda s, N}(x))$ is uniformly bounded with respect to $x \in X$ and $s \in (0, 1]$.   □

Proof of Theorem 3.3.7 Fix $\lambda > 1$. By Proposition 3.1.16, there exists $c \geq 1$ such that $cg(w) \geq s \geq g(w)$ if $w \in \Lambda_s$. Since g is sub-exponential, there exist $c_1 \in (0, 1)$ and $m \geq 1$ such that $c_1 g(w) \geq g(v)$ whenever $v \in T_w$ and $|v| \geq |w| + m$. Assume that $w \in \Lambda_s$. Set $w_* = (w)_{\lambda s}$. Then $\lambda s \geq g(w_*)$. If $|w| \geq |w_*| + nm$, then $(c_1)^n g(w_*) \geq g(w)$ and hence $(c_1)^n \lambda s \geq g(w) \geq s/c$. This shows that $(c_1)^n \lambda c \geq 1$. Set $l = \min\{n | n \geq 0, (c_1)^n \lambda c < 1\}$. Then we see that $|w| < |w_*| + lm$.

On the other hand, since μ is super-exponential, there exists $c_2 > 0$ such that $\mu(K_u) \geq c_2 \mu(K_{\pi(u)})$ for any $u \in T$. This implies that $\mu(K_{w_*}) \leq (c_2)^{-ml} \mu(K_w)$. Since μ is gentle, there exists $c_* > 0$ such that $\mu(K_{w(1)}) \leq c_* \mu(K_{w(2)})$ whenever $w(1), w(2) \in \Lambda_s$ and $K_{w(1)} \cap K_{w(2)} \neq \emptyset$ for some $s \in (0, 1]$. Therefore for any $u \in \Lambda_{\lambda s, 2M+1}(w_*)$,

$$\mu(K_u) \leq (c_*)^{2M+1} \mu(K_{w_*}) \leq (c_*)^{2M+1} (c_2)^{-ml} \mu(K_w).$$

Thus we have shown that

$$\Lambda_s = \Theta(s, \lambda, 2M + 1, (c_*)^{2M+1} (c_2)^{-ml})$$

for any $s \in (0, 1]$. Now by Lemma 3.3.8, μ has M-volume doubling property with respect to g for any $M \geq 1$.   □

In order to study the converse direction of Theorem 3.3.7, we need the thickness of K with respect to the weight function in question.

Theorem 3.3.9 *Let $g : T \to (0, 1]$ be a weight function and let $\mu \in \mathcal{M}_P(X, \mathcal{O})$. Assume that g is thick.*

(1) *Suppose that g is exponential and uniformly finite. Then the following conditions are equivalent:*

   (VD1)   *μ has M-volume doubling property with respect to g for some $M \geq 1$.*
   (VD2)   *μ has M-volume doubling property with respect to g for any $M \geq 1$.*
   (VD3)   *μ is gentle with respect to g and μ is super-exponential.*

(2) *Suppose that K is minimal and g is super-exponential and that $\#(S(w)) \geq 2$ for any $w \in T$. Then (VD1), (VD2) and the following condition (VD4) are equivalent:*

   (VD4)   *g is sub-exponential and uniformly finite, μ is gentle with respect to g and μ is super-exponential.*

Moreover, if any of the above conditions (VD1), (VD2) *and* (VD4) *hold, then μ is exponential and*

$$\sup_{w \in T} \#(S(w)) < +\infty.$$

In general, the statement of Theorem 3.3.9 is false if g is not thick. In fact, in Example 3.4.10, we will present an example without thickness where d is adapted to g, g is exponential and uniformly finite, μ has the volume doubling property with respect to g but μ is neither gentle to g nor super-exponential.

As for a proof of Theorem 3.3.9, it is enough to show the following theorem.

Theorem 3.3.10 *Let $g : T \to (0, 1]$ be a weight function and let $\mu \in \mathcal{M}_P(X, \mathcal{O})$. Assume that μ has M-volume doubling property with respect to g for some $M \geq 1$.*

(1) *If g is thick, then μ is gentle with respect to g.*
(2) *If g is thick and g is super-exponential, then μ is super-exponential.*
(3) *If g is thick and K is minimal, then g is uniformly finite.*
(4) *If g is thick, K is minimal, μ is super-exponential, then*

$$\sup_{w \in T} \#(S(w)) < +\infty.$$

Moreover, if $\#(S(w)) \geq 2$ for any $w \in T$ in addition, then μ is sub-exponential.
(5) *If g is uniformly finite, μ is gentle with respect to g, μ is sub-exponential, then g is sub-exponential.*

To prove Theorem 3.3.10, we need several lemmas.

Lemma 3.3.11 *Let $g : T \to (0, 1]$ be a weight function. Assume that K is minimal and g is thick. Let $\mu \in \mathcal{M}_P(X, \mathcal{O})$. If μ has M-volume doubling property with respect to g for some $M \geq 1$, then there exists $c > 0$ such that $\mu(O_w) \geq c\mu(K_w)$ for any $w \in T$.*

Proof By Proposition 3.2.2, there exists $\gamma > 0$ such that $O_v \supseteq U_M(x, \gamma s)$ for some $x \in K_v$ if $v \in \Lambda_s$. Let $w \in T$. Choose $u \in T_w$ such that $u \in \Lambda_{g(w)/2}$. Then

$$\mu(O_w) \geq \mu(O_u) \geq \mu(U_M(x, \gamma g(w)/2)).$$

Since μ has M-volume doubling property with respect to g, there exists $c > 0$ such that

$$\mu(U_M(y, \gamma r/2)) \geq c\mu(U_M(y, r))$$

for any $y \in X$ and $r > 0$. Since $U_M(x, g(w)) \supseteq K_w$, it follows that

$$\mu(O_w) \geq \mu(U_M(x, \gamma g(w)/2)) \geq c\mu(U_M(x, g(w))) \geq c\mu(K_w).$$

$\square$

Lemma 3.3.12 *Let* $g : T \to (0, 1]$ *be a weight function. Assume that* $\mu \in \mathcal{M}_P(X, \mathcal{O})$ *is gentle with respect to* g *and that* g *is uniformly finite. Then there exists* $c > 0$ *such that*

$$c\mu(K_w) \geq \mu(U_M(x, s))$$

if $w \in \Lambda_{s,0}(x)$.

Proof Since μ is gentle with respect to g, there exists $c_1 > 0$ such that $\mu(K_v) \leq c_1\mu(K_w)$ if $w \in \Lambda_s$ and $v \in \Lambda_{s,1}(w)$. Hence if $v \in \Lambda_{s,M+1}(w)$, then it follows that $\mu(K_v) \leq (c_1)^{M+1}\mu(K_w)$. Since $\Lambda_{s,M}(x) \subseteq \Lambda_{s,M+1}(w)$,

$$\mu(U_M(x, s)) \leq \sum_{v \in \Lambda_{s,M}(x)} \mu(K_v)$$

$$\leq \sum_{v \in \Lambda_{s,M}(x)} (c_1)^{M+1}\mu(K_w) = (c_1)^{M+1}\#(\Lambda_{s,M}(x))\mu(K_w).$$

By Lemma 3.1.17, we obtain the desired statement. $\square$

Proof of Theorem 3.3.10 (1) Since g is thick, there exists $\beta \in (0, 1)$ such that, for any $s \in (0, 1]$ and $w \in \Lambda_s$, $K_w \supseteq U_M(x, \beta s)$ for some $x \in K_w$. By M-volume doubling property of μ, there exists $c > 0$ such that $\mu(U_M(x, \beta s)) \geq c\mu(U_M(x, s))$ for any $s \in (0, 1]$ and $x \in X$. Hence

$$\mu(K_w) \geq \mu(U_M(x, \beta s)) \geq c\mu(U_M(x, s)). \tag{3.3.1}$$

If $v \in \Lambda_s$ and $K_v \cap K_w \neq \emptyset$, then $U_M(x, s) \supseteq K_v$. (3.3.1) shows that $\mu(K_w) \geq c\mu(K_v)$. Hence μ is gentle with respect to g.

(2) Let $v \in T \setminus \{\phi\}$. Choose $u \in T_v$ so that $u \in \Lambda_{g(v)/2}$. Applying (3.3.1) to u and using the volume doubling property repeatedly, we see that there exists $x \in K_u$ such that

$$\mu(K_v) \geq \mu(K_u) \geq \mu(U_M(x, \beta g(v)/2)) \geq c^n\mu(U_M(x, \beta^{1-n}g(v)/2)) \tag{3.3.2}$$

for any $n \geq 0$. Since g is super-exponential, there exists $n \geq 0$, which is independent of v, such that $\beta^{1-n}g(v)/2 > g(\pi(v))$. By (3.3.2), we obtain $\mu(K_v) \geq c^n\mu(K_{\pi(v)})$. Thus μ is super-exponential.

(3) Let $w \in \Lambda_s$. Then $\{O_v\}_{v \in \Lambda_{s,1}(w)}$ is mutually disjoint by Lemma 2.2.2-(2). By (3.3.1) and Lemma 3.3.11,

$$\mu(K_w) \geq c\mu(U_M(x, s)) \geq c \sum_{v \in \Lambda_{s,1}(w)} \mu(O_v) \geq c' \sum_{v \in \Lambda_{s,1}(w)} \mu(K_v).$$

As μ is gentle with respect to g by (1), there exists $c_* > 0$, which is independent of w and s, such that $\mu(K_v) \geq c_*\mu(K_w)$ for any $v \in \Lambda_{s,1}(w)$. Therefore,

$$\mu(K_w) \geq c' \sum_{v \in \Lambda_{s,1}(w)} \mu(K_v) \geq c'c_*\#(\Lambda_{s,1}(w))\mu(K_w).$$

Hence $\#(\Lambda_{s,1}(w)) \leq (c'c_*)^{-1}$ and g is uniformly finite.

(4) By Lemma 3.3.11, for any $w \in T$, we have

$$\mu(K_w) \geq \mu(\cup_{v \in S(w)} O_v) = \sum_{v \in S(w)} \mu(O_v) \geq c \sum_{v \in S(w)} \mu(K_v).$$

Since μ is super-exponential, there exists $c' > 0$ such that $\mu(K_v) \geq c'\mu(K_w)$ if $w \in T$ and $v \in S(w)$. Hence

$$\mu(K_w) \geq c \sum_{v \in S(w)} \mu(K_v) \geq cc'\#(S(w))\mu(K_w).$$

Thus $\#(S(w)) \leq (cc')^{-1}$, which is independent of w.

Next assume that $\#(S(w)) \geq 2$ for any $w \in T$. By the above arguments,

$$\mu(O_v) \geq c\mu(K_v) \geq c_*\mu(K_w) \geq c_*\mu(O_w) \tag{3.3.3}$$

for any $w \in T$ and $v \in S(w)$, where $c_* = cc'$. Let $v_* \in S(w)$. If $\mu(O_{v_*}) = (1-a)\mu(O_w)$, then

$$\mu(O_w) \geq \sum_{v \in S(w)} \mu(O_v) = (1-a)\mu(O_w) + \sum_{v \in S(w), v \neq v_*} \mu(O_v).$$

This implies $a\mu(O_w) \geq \mu(O_v)$ for any $v \in S(w)\backslash\{v_*\} \neq \emptyset$. By (3.3.3), $a \geq c_*$. Therefore, $\mu(O_v) \leq (1-c_*)\mu(O_w)$ for any $v \in S(w)$. This implies

$$c\mu(K_v) \leq \mu(O_v) \leq (1-c_*)^m\mu(O_w) \leq (1-c_*)^m\mu(K_w)$$

if $v \in T_w$ and $|v| = |w| + m$. Choosing m so that $(1-c_*)^m < c$, we see that μ is sub-exponential.

(5) As μ is sub-exponential, there exist $\alpha \in (0,1)$ and $m \geq 0$ such that $\mu(K_v) \leq \alpha\mu(K_w)$ if $u \in T_w$ and $|u| \geq |w| + m$. Since μ has M-volume doubling property with respect to g, there exist $\lambda, c \in (0,1)$ such that $\mu(U_M(x, \lambda s)) \geq c\mu(U_M(x, s))$ for any $x \in X$ and $s > 0$. Let $\beta \in (\lambda, 1)$. Assume that g is not sub-exponential. Then for any $n \geq 0$, there exist $w \in T$ and $u \in T_w$ such that $|u| \geq |w| + nm$ and $g(u) \geq \beta g(w)$. Set $s = g(w)$. Since $\beta > \lambda$, there exists $u_* \in T_u \cap \Lambda_{\lambda s}$. Let $x \in K_{u_*}$. Then by the volume doubling property,

$$\mu(U_M(x, \lambda s)) \geq c\mu(U_M(x, s)) \geq c\mu(K_w).$$

By Lemma 3.3.12, there exists $c_* > 0$, which is independent of n, w and u, such that

$$c_* \mu(K_{u_*}) \geq \mu(U_M(x, \lambda s)).$$

Since μ is sub-exponential,

$$\alpha^n c_* \mu(K_w) \geq c_* \mu(K_{u_*}) \geq \mu(U_M(x, \lambda s)) \geq c\mu(K_w).$$

This implies $\alpha^n c_* \geq c$ for any $n \geq 0$ which is a contradiction.   □

At the end of this section, we give a proof of Theorem 3.1.21 by using Theorem 3.3.10.

Proof of Theorem 3.1.21 Assume that $(g_d)^\alpha \underset{\mathrm{BL}}{\sim} g_\mu$ and d is uniformly finite. Since d is adapted, there exist $M \geq 1$ and $\alpha_1, \alpha_2 > 0$ such that

$$U_M^d(x, \alpha_1 r) \subseteq B_d(x, r) \subseteq U_M^d(x, \alpha_2 r)$$

for any $x \in X$ and $r > 0$.

Write $d_w = g_d(w)$ and $\mu_w = g_\mu(w)$. Since $(g_d)^\alpha \underset{\mathrm{BL}}{\sim} g_\mu$, there exist $c_1, c_2 > 0$ such that

$$c_1 (d_w)^\alpha \leq \mu_w \leq c_2 (d_w)^\alpha$$

for any $w \in T$. For any $x \in X$, choose $w \in T$ so that $x \in K_w$ and $w \in \Lambda_{\alpha_1 r}^d$. Since d is super-exponential, there exists $\lambda > 0$ such that $d_w \geq \lambda d_{\pi(w)}$ for any $w \in T$. Then

$$\mu(B_d(x, r)) \geq \mu(U_M^d(x, \alpha_1 r)) \geq \mu(K_w) \geq c_1(d_w)^\alpha \geq c_1(\lambda d_{\pi(w)})^\alpha \geq c_1(\lambda\alpha_1 r)^\alpha.$$

On the other hand, since d is uniformly finite, Lemma 3.1.17 implies

$$\mu(B_d(x, r)) \leq \mu(U_M^d(x, \alpha_2 r)) \leq \sum_{w \in \Lambda_{\alpha_2 r, M}^d(x)} \mu(K_w)$$

$$\leq (c_2)^\alpha \sum_{w \in \Lambda_{\alpha_2 r, M}^d(x)} (d_w)^\alpha \leq (c_2)^\alpha \#(\Lambda_{\alpha_2 r, M}^d(x))(\alpha_2 r)^\alpha \leq C_2 r^\alpha,$$

where $C_2 > 0$ is independent of x and r. Conversely, assume (3.1.7). For any $w \in T$ and $x \in K_w$,

$$K_w \subseteq U_M^d(x, d_w) \subseteq B_d(x, d_w/\alpha_1).$$

Hence

$$\mu(K_w) \le \mu(B_d(x, d_w/\alpha_1)) \le C_2(d_w/\alpha_1)^\alpha.$$

By Proposition 3.2.1, there exists $z \in K_w$ such that

$$K_w \supseteq U_M^d(z, \beta d_{\pi(w)}) \supseteq B_d(z, \beta d_{\pi(w)}/\alpha_2).$$

By (3.1.7),

$$\mu(K_w) \ge \mu(B_d(z, \beta d_{\pi(w)}/\alpha_2)) \ge C_1(\beta d_{\pi(w)}/\alpha_2)^\alpha \ge C_1(\beta d_w/\alpha_2)^\alpha.$$

Thus we have shown that $g_d \underset{\text{BL}}{\sim} g_\mu$. Furthermore, since d is M-adapted for some $M \ge 1$, μ has M-volume doubling property with respect to the weight function g_d. Applying Theorem 3.3.10-(3), we see that g_d is uniformly finite. In the same way, by Theorem 3.3.10, both g_d and g_μ are exponential. □

3.4 Example: Subsets of the Square

In this section, we give illustrative examples of the results in the previous sections. For simplicity, our examples are subsets of the square $[0, 1]^2$, denoted by Q, and trees parametrizing partitions are sub-trees of $(T^{(9)}, \mathcal{A}^{(9)}, \phi)$ defined in Example 2.1.3. Note that $[0, 1]^2$ is divided into 9-squares with the length of the sides $\frac{1}{3}$. As in Example 2.2.7, the tree $(T^{(9)}, \mathcal{A}^{(9)}, \phi)$ appears as the tree parametrizing the natural partition associated with this self-similar division. Namely, let $p_1 = (0, 0)$, $p_2 = (1/2, 0)$, $p_3 = (1, 0)$, $p_4 = (1, 1/2)$, $p_5 = (1, 1)$, $p_6 = (1/2, 1)$, $p_7 = (0, 1)$, $p_8 = (0, 1/2)$ and $p_9 = (1/2, 1/2)$. Set $W = \{1, \ldots, 9\}$. Define $F_i : Q \to Q$ by

$$F_i(x) = \frac{1}{3}(x - p_i) + p_i$$

for any $i \in W$. Then F_i is a similitude for any $i \in W$ and

$$Q = \bigcup_{i \in W} F_i(Q).$$

See Fig. 3.1. In this section, we write $(W_*, \mathcal{A}_*, \phi) = (T^{(9)}, \mathcal{A}^{(9)}, \phi)$, which is a locally finite tree with a reference point ϕ. Set $W_m = \{1, \ldots, 9\}^m$. Then $(W_*)_m = W_m$ and $\pi^{(W_*, \mathcal{A}_*, \phi)}(w) = w_1 \ldots w_{m-1}$ for any $w = w_1 \ldots w_m \in W_m$. For simplicity, we use $|w|$ and π in place of $|w|_{(W_*, \mathcal{A}_*, \phi)}$ and $\pi^{(W_*, \mathcal{A}_*, \phi)}$ respectively hereafter. Define $g : W_* \to (0, 1]$ by $g(w) = 3^{-|w|}$ for any $w \in W_*$. Then g is an exponential weight function.

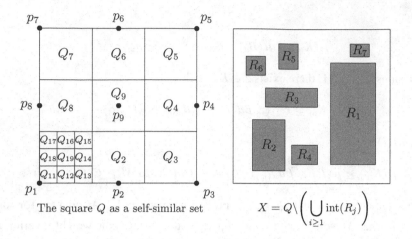

Fig. 3.1 The square Q and its subset X

As for the natural associated partition of Q, define $F_w = F_{w_1} \circ \ldots \circ F_{w_m}$ and $Q_w = F_w(Q)$ for any $w = w_1 \ldots w_m \in W_m$. Set $Q_*(w) = Q_w$ for any $w \in W_*$. (If $w = \phi$, then F_ϕ is the identity map and $Q_\phi = Q$.) Then $Q_* : W_* \to \mathcal{C}(Q, \mathcal{O})$ is a partition of Q parametrized by $(W_*, \mathcal{A}_*, \phi)$, where $\mathcal{O}$ is the natural topology induced by the Euclidean metric. In fact, $\cap_{m \geq 0} Q_{[\omega]_m}$ for any $\omega \in \Sigma$, where $\Sigma = W^{\mathbb{N}}$, is a single point. Define $\sigma : \Sigma \to Q$ by $\{\sigma(\omega)\} = \cap_{m \geq 0} Q_{[\omega]_m}$.

It is easy to see that the partition Q_* is minimal, g is uniformly finite, g is thick with respect to the partition Q_*, and the (restriction of) Euclidean metric d_E on Q is 1-adapted to g.

In order to have more interesting examples, we consider certain class of subsets of Q whose partition is parametrized by a subtree $(T, \mathcal{A}_*|_{T \times T}, \phi)$ of $(W_*, \mathcal{A}_*, \phi)$. Let $\{I_m\}_{m \geq 0}$ be a sequence of subsets of W_* satisfying the following conditions (SQ1), (SQ2) and (SQ3):

(SQ1)   For any $m \geq 0$, $I_m \subseteq W_m$ and if $\widehat{I_{m+1}} = \{wi \,|\, w \in I_m, i \in W\}$, then $I_{m+1} \supseteq \widehat{I_{m+1}}$.

(SQ2) $Q_w \cap Q_v = \emptyset$ if $w \in \widehat{I_{m+1}}$ and $v \in I_{m+1} \backslash \widehat{I_{m+1}}$.

(SQ3)   For any $m \geq 0$, the set $\cup_{w \in I_m} Q_w$ is a disjoint union of rectangles $R_j^m = [a_j^m, b_j^m] \times [c_j^m, d_j^m]$ for $j = 1, \ldots, k_m$.

See Fig. 3.1. By (SQ2), we may assume that $k_m \leq k_{m+1}$ and $R_j^m = R_j^{m+1}$ for any m and $j = 1, \ldots, k_m$ without loss of generality. Under this assumption, we may omit m of $R_j^m, a_j^m, b_j^m, c_j^m$ and d_j^m and simply write R_j, a_j, b_j, c_j and d_j respectively.

Notation We use int(A) and ∂A to denote the interior and the boundary, respectively, of a subset A of Q with respect to the natural topology $\mathcal{O}$ induced by the Euclidean metric.

Note that $\mathrm{int}(\cup_{w \in I_m} Q_w) = \cup_{j=1,\dots,k_m} \mathrm{int}(R_j)$.

Proposition 3.4.1

(1) *Define*

$$X^{(m)} = Q \Big\backslash \Big(\bigcup_{j=1,\dots,k_m} \mathrm{int}(R_j) \Big).$$

then $X^{(m)} \supseteq X^{(m+1)}$ for any $m \geq 0$ and $X = \cap_{m \geq 0} X^{(m)}$ is a non-empty compact set. Moreover, $\partial R_j \subseteq X$ for any $j \geq 1$.

(2) *Define $(T)_m = \{w \mid w \in W_m, \mathrm{int}(Q_w) \cap X \neq \emptyset\}$ for any $m \geq 0$. If $T = \cup_{m \geq 0}(T)_m$ and $\mathcal{A} = \mathcal{A}_* |_{T \times T}$, then $(T, \mathcal{A}, \phi)$ is a locally finite tree with the reference point ϕ and $\#(S(w)) \geq 3$ for any $w \in T$. Moreover, let*

$$\Sigma_T = \{\omega \mid \omega \in \Sigma, [\omega]_m \in (T)_m \text{ for any } m \geq 0\}$$

Then $X = \sigma(\Sigma_T)$.

(3) *Define $K_w = Q_w \cap X$ for any $w \in T$. Then $K_w \neq \emptyset$ and $K : T \to \mathcal{C}(X)$ defined by $K(w) = K_w$ is a minimal partition of X parametrized by $(T, \mathcal{A}, \phi)$. Moreover, $g|_T$ is exponential and uniformly finite.*

To prove the above proposition, we need the following lemma.

Lemma 3.4.2 *If $w \in T$, then $\cup_{i \in W, wi \notin T} Q_{wi}$ is a disjoint union of rectangles and $\#(\{i \mid i \in W, wi \in T\}) \geq 3$.*

Proof Set $I = \{i \mid i \in W, wi \notin T\}$. For each $i \in I$, there exists $n_i \geq 1$ such that $Q_{wi} \subseteq R_{n_i}$. Hence $\cup_{i \in I} Q_{wi} = \cup_{i \in I}(Q_w \cap R_{n_i})$. Since $\{R_j\}_{j \geq 1}$ are mutually disjoint, $\cup_{i \in I} Q_{wi}$ is a disjoint union of rectangles. If $Q_{wi} \cap Q_{wj} \neq \emptyset$ for $i \neq j \in I$, then $R_{n_i} \cap R_{n_j} \neq \emptyset$ and hence $R_{n_i} = R_{n_j}$. Therefore say $1, 9 \in I$ for example. Then $R_{n_1} = R_{n_9}$. Since R_{n_1} is a rectangle, it follows that $R_{n_1} \supseteq Q_{w2} \cup Q_{w8}$ and so $2, 8 \in I$ as well. By such an argument, if $\#(I) \geq 7$, then R_{n_i}'s are the same for any $i \in I$ and $R_{n_i} \supseteq Q_w$ for any $i \in I$. This contradicts the fact that $w \in T$. Thus we have $\#(I) \leq 6$.   □

Proof of Proposition 3.4.1 (1) Since $\{X^{(m)}\}_{m \geq 0}$ is a decreasing sequence of compact sets, X is a nonempty compact set. By (SQ2), $R_j \cap R_i = \emptyset$ for any $i \neq j$. Therefore, $\partial R_j \subseteq X^{(m)}$ for any $m \geq 0$. Hence $\partial R_j \subseteq X$.

(2) If $w \in (T)_m$, then $\mathrm{int}(Q_{\pi(w)}) \cap X \supseteq \mathrm{int}(Q_w) \cap X \neq \emptyset$. Hence $\pi(w) \in (T)_{m-1}$. Using this inductively, we see that $[w]_k \in (T)_k$ for any $k \in \{0, 1, \dots, m\}$. This implies that $(T, \mathcal{A}, \phi)$ is a locally finite tree with a reference point ϕ. By Lemma 3.4.2, we see that $\#(\{i \mid i \in W, wi \in (T)_{m+1}\}) \geq 3$. Next if $\omega \in \Sigma_T$, then for any $m \geq 0$, there exists $x_m \in \mathrm{int}(Q_{[\omega]_m}) \cap X$. Therefore, $x_m \to \sigma(\omega)$ as $m \to \infty$. Since X is compact, it follows that $\sigma(\omega) \in X$.

Conversely, assume that $x \in X$. Set $W_{m,x} = \{w | w \in W_m, x \in Q_w\}$. Note that $\#(\sigma^{-1}(x)) \le 4$ and $\cup_{w \in W_{m,x}} Q_w$ is a neighborhood of x. Suppose that $(T)_m \cap W_{m,x} \ne \emptyset$ for any $m \ge 0$. Then there exists $w_m \in (T)_m \cap W_{m,x}$ such that $x \in Q_{w_m}$. Since $W_{m,x} = \{[\omega]_m | \omega \in \sigma^{-1}(x)\}$, there exists $\omega \in \sigma^{-1}(x)$ such that $[\omega]_m = w_m$ for infinitely many m. As $\text{int}(Q_{[\omega]_m})$ is monotonically decreasing, it follows that $[\omega]_m \in (T)_m$ for any $m \ge 0$. Hence $\omega \in \Sigma_T$ and $\sigma(\omega) = x$. Suppose that there exists $m \ge 0$ such that $W_{m,x} \cap (T)_m = \emptyset$. By this assumption, $\text{int}(Q_w) \cap X = \emptyset$ for any $w \in W_{m,x}$ and hence there exists $j_w \ge 1$ such that $Q_w \subseteq R_{j_w}$. Note that $Q_w \cap Q_{w'} \ne \emptyset$ for any $w, w' \in W_{m,x}$ and hence $R_{j_w} = R_{j_{w'}}$. Therefore, $\cup_{w \in W_{m,x}} Q_w \subseteq R_j$ for some $j \ge 1$. Since $x \in \text{int}(\cup_{w \in W_{m,x}} Q_w)$, it follows that $x \notin X$. This contradiction concludes the proof of (2).

(3) The fact that K is a partition of X parametrized by $(T, \mathcal{A}|_{T \times T}, \phi)$ is straightforward from (1) and (2). As $K_w \backslash (\cup_{v \in (T)_m, v \ne w} K_v)$ is contained in the sides of the square Q_w, the partition K is minimal. Since g is exponential, $g|_T$ is exponential as well. Furthermore, $\Lambda_{s,1}^{g|_T}(w) \subseteq \{v | v \in W_m, Q_v \cap Q_w \ne \emptyset\}$ for any $w \in (T)_m$. Hence $\#(\Lambda_{s,1}^{g|_T}(w)) \le 9$. This shows that $g|_T$ is uniformly finite. $\square$

Now, we consider when the restriction of the Euclidean metric is adapted.

Definition 3.4.3 Let $R = [a, b] \times [c, d]$ be a rectangle. The degree of distortion of R, $\kappa(R)$, is defined by

$$\kappa(R) = \max \left\{ 1, (1 - \delta_{c0})(1 - \delta_{d1}) \frac{|b - a|}{|d - c|}, (1 - \delta_{a0})(1 - \delta_{b1}) \frac{|d - c|}{|b - a|} \right\},$$

where δ_{xy} is the Kronecker delta defined by $\delta_{xy} = 1$ if $x = y$ and $\delta_{xy} = 0$ if $x \ne y$. Moreover, for $\kappa \ge 1$, we define

$$\mathcal{R}_\kappa^0 = \{R | R \text{ is a rectangle}, R \subseteq Q \text{ and } \kappa(R) \le \kappa\}$$

and

$$\mathcal{R}_\kappa^1 = \{R | R \subseteq Q, R \text{ is a rectangle, there exists } w \in T \text{ such that } Q_w \backslash \text{int}(R)$$

$$\text{has two connected components and } \kappa(Q_w \cap R) \le \kappa\}.$$

The extra factors $(1 - \delta_{c0})$, $(1 - \delta_{d1})$, $(1 - \delta_{a0})$ and $(1 - \delta_{b1})$ become effective if the rectangle R has an intersection with the boundary of the square Q.

Theorem 3.4.4 *Let d be the restriction of the Euclidean metric on X. Then d is adapted to $g|_T$ if and only if the following condition (SQ4) holds:*

(SQ4)   *There exists $\kappa \geq 1$ such that $R_j \in \mathcal{R}_\kappa^0 \cup \mathcal{R}_\kappa^1$ for any $j \geq 1$.*

Several lemmas are needed to prove the above theorem.

If $w, v \in (T)_m$ and $Q_w \cap Q_v = \emptyset$, then $\min_{x \in Q_w, y \in Q_v} d(x, y) \geq 3^{-m}$. Hence if $x, y \in X$ and $d(x, y) < 3^{-m}$, then there exist $u, v \in (T)_m$ such that $x \in Q_w$, $y \in Q_v$ and $Q_w \cap Q_v \neq \emptyset$.

Lemma 3.4.5 *Define $N(x, y) = [-\frac{\log d(x,y)}{\log 3}]$ for any $x, y \in X$, where $[a]$ is the integer part of a real number a. Then,*

$$\frac{1}{3^{N(x,y)+1}} < d(x, y) \leq \frac{1}{3^{N(x,y)}} \tag{3.4.1}$$

for any $x, y \in X$ and there exist $w, v \in (T)_{N-1}$ such that $x \in Q_w$, $y \in Q_v$ and $Q_w \cap Q_v \neq \emptyset$.

Proof (3.4.1) is immediate. If $w, v \in (T)_m$ and $Q_w \cap Q_v = \emptyset$, then

$$\min_{x \in Q_w, y \in Q_v} d(x, y) \geq 3^{-m}.$$

Hence if $x, y \in X$ and $d(x, y) < 3^{-m}$, then there exist $u, v \in (T)_m$ such that $x \in Q_w$, $y \in Q_v$ and $Q_w \cap Q_v \neq \emptyset$. This yields the rest of the statement. □

Notation For integers $n, k, l \geq 0$, we set

$$Q(n, k, l) = \left[\frac{k}{3^n}, \frac{(k+1)}{3^n}\right] \times \left[\frac{l}{3^n}, \frac{(l+1)}{3^n}\right].$$

Lemma 3.4.6 *Assume (SQ4). Let $M = [\log(2\kappa)/\log 3] + 1$ and $L = 2[2\kappa + 1]$. If $w, v \in (T)_m$, $w \neq v$ and $Q_w \cap Q_v \neq \emptyset$, then there exists a chain $(w(1), \ldots, w(p))$ of K such that $w \in T_{w(1)}$, $v \in T_{w(p)}$, $|w(i)| \geq m - M$ for any $i = 1, \ldots, p$ and $p \leq L + 1$.*

Proof Case 1: Assume that $Q_w \cap Q_v$ is a line segment. Without loss of generality, we may assume that $Q_w = Q(m, k - 1, l)$ and $Q_v = Q(m, k, l)$.

Case 1a: $K_w \cap K_v \neq \emptyset$, then (w, v) is a desired chain of K.

Case 1b: In case $K_w \cap K_v = \emptyset$, $Q_w \cap Q_v \cap K_w$ and $Q_w \cap Q_v \cap K_v$ are disjoint closed subsets of $Q_w \cap Q_v$. Since $Q_w \cap Q_v$ is connected, there exists $a \in Q_w \cap Q_v$ such that $a \notin K_w \cup K_v$. Since $K_w \cup K_v$ is closed, there exists an open neighborhood of a which has no intersection with $K_w \cup K_v$. This open neighborhood must be contained in $R_j = [a_j, b_j] \times [c_j, d_j]$ for some j. So, we see that $R_j \cap \text{int}(Q_w \cap$

$Q_v) \neq \emptyset$ and $a_j < k/3^m < b_j$. Since both $[a_j, b_j] \times \{c_j\}$ and $[a_j, b_j] \times \{d_j\}$ are contained in X and $K_w \cap K_v \neq \emptyset$, we see that $c_j < l/3^m$ and $d_j > (l+1)/3^j$. Since neither $\text{int}(Q_w)$ nor $\text{int}(Q_v)$ is contained in R_j, it follows that $(k-1)/3^m < a_j$ and $b_j < (k+1)/3^m$. Now if $R_j \in \mathcal{R}_\kappa^0$, it follows that $|d_j - c_j| \leq \kappa |a_j - b_j| \leq 2\kappa/3^m$. Choose n so that $(l+n)/3^m < d_j \leq (l+n+1)/3^m$. Then for each $i = 1, \ldots, n+1$, there exist $w_*(i), v_*(i) \in (T)_m$ such that $Q_{w_*(i)} = Q(m, k-1, l+i-1)$ and $Q_{v_*(i)} = Q(m, k, l+i-1)$. Then $\mathbf{w} = (w_*(1), \ldots, w_*(n+1), v_*(n+1), \ldots, v_*(1))$ is a chain of K, $w_*(1) = w$ and $v_*(1) = v$. Since $(n-1)/3^m \leq |c_j - d_j| \leq 2\kappa/3^m$, we see that $n \leq [2\kappa + 1]$. Hence $2n \leq 2[2\kappa+1] \leq L$. Therefore $\mathbf{w}$ is a desired chain. Next assume $R_j \in \mathcal{R}_\kappa^1$. By the definition of $\mathcal{R}_\kappa^1$, there exists $u \in (T)_{m-q}$ such that $Q_u \backslash R_j$ has two connected component and $k(Q_u \cap R_j) = 3^{-m+q}/|a_j - b_j| \leq \kappa$. Hence $3^{-m+q} \leq 2\kappa/3^m$ and so $3^q \leq 2\kappa$. This shows $q \leq M$. Sifting Q_u in the vertical direction, we may find $u' \in (T)_{m-M}$ such that $Q_w \cup Q_v \subseteq Q_{u'}$. Then (u') is a desired chain.

Case 2: Assume that $Q_w \cap Q_v$ is a single point. Without loss of generality, we may assume that $Q_w = Q(m, k-1, l-1)$ and $Q_v = Q(m, k, l)$. Choose $u(1), u(2) \in W_m$ so that $Q_{u(1)} = Q(m, k-1, l)$ and $Q_{w(2)} = Q(m, k+1, l-1)$. If neither $u(1)$ nor $u(2)$ belongs to T, then there exist $i, j \geq 1$ such that $Q_{u(1)} \subseteq R_i$ and $Q_{u(2)} \subseteq R_j$. Since $Q_{u(1)} \cap Q_{u(2)} \neq \emptyset$, it follows that $R_i = R_j$ and hence $Q_w \cup Q_v \subseteq R_i$. This contradicts the fact that $w, v \in T$. Hence $u(1) \in T$ or $u(2) \in T$. Let $u(1) \in T$. Then $Q_w \cap Q_{u(1)}$ and $Q_{u(1)} \cap Q_v$ are line segments. If $K_w \cap K_{u(1)} = \emptyset$ and $K_{u(1)} \cap K_v = \emptyset$, then we have rectangles R_i and R_j which are constructed by the arguments in (1). Since those rectangles have a common point $(k/3^m, l/3^l)$, we see that $R_j = R_i$ and hence it covers $Q_{u(1)}$. This contradicts the fact that $u(1) \in T$. Therefore, either $K_w \cap K_{u(1)}$ or $K_{u(1)} \cap K_v$ is not empty. Now combining two chains for $(w, u(1))$ and $(u(1), v)$ constructed in (1), we obtain a desired chain.   □

Proof of Theorem 3.4.4 Assume (SQ4). Let $x, y \in X$. Define $N = N(x, y)$ and choose $w, v \in W_{N-1}$ as in Lemma 3.4.5. We fix the constants M and L as in Lemma 3.4.6. Then since $Q_w \cap Q_v \neq \emptyset$, Lemma 3.4.6 implies the existence of a chain $(w(1), \ldots, w(p)) \in \mathcal{CH}_K(x, y)$ satisfying $p \leq L+1$ and $N-1 \geq |w(i)| \geq N-1-M$. By (3.4.1),

$$3^{M+2} d(x, y) \geq \frac{1}{3^{(N-M-1)}} \geq \frac{1}{3^{|w(i)|}} = g(w(i)).$$

Thus we have verified the conditions (ADb)$_L$ in Theorem 2.4.5. Since (ADa) is obvious, d is L-adapted to $g|_T$ by Theorem 2.4.5.

Conversely, assume that d is J-adapted to $g|_T$. By (ADb)$_J$, there exists $C \geq 0$ such that for any $x, y \in X$, there exists a chain $(w(1), \ldots, w(J+1)) \in \mathcal{CH}_K(x, y)$ satisfying

$$Cd(x, y) \geq \frac{1}{3^{|w(i)|}} \tag{3.4.2}$$

for any $i = 1, \ldots, J+1$. Set $M_* = [\log C / \log 3] + 1$. Suppose that (SQ4) does not hold; for any $\kappa \geq 1$, there exists $R_j \notin \mathcal{R}_\kappa^0 \cup \mathcal{R}_\kappa^1$. In particular, we choose $\kappa > 3^{M_*}$. Write $R = R_j$ and set $R = [a, b] \times [c, d]$. Define $\partial R_L = \{a\} \times [c, d]$ and $\partial R_R = \{b\} \times [c, d]$. (The symbols "L" and "R" correspond to the words "Left" and "Right" respectively.) Without loss of generality, we may assume that $|a - b| \leq |c - d|$. Since $R \notin \mathcal{R}_\kappa^0$, we have $\kappa|b - a| \leq |d - c|$. Let $x = (a, (c + d)/2)$ and let $y = (b, (c + d)/2)$. Set $N = N(x, y)$. There exists $(w(1), \ldots, w(J+1)) \in \mathcal{CH}_K(x, y)$ such that (3.4.2) holds for any $i = 1, \ldots, J + 1$. By (3.4.1),

$$|w(i)| \geq N - M_*, \tag{3.4.3}$$

for any $i = 1, \ldots, J+1$. Define $A = [0, 1] \times (c, d)$. If $Q_{w(i)} \subseteq A$, $Q_{w(i)} \cap \partial R_L \neq \emptyset$ and $Q_{w(i)} \cap \partial R_R \neq \emptyset$, then the fact that $R \notin \mathcal{R}_\kappa^1$ along with (3.4.1) shows

$$\frac{1}{3^{|w(i)|}} \geq \kappa|b - a| = \kappa d(x, y) \geq \frac{\kappa}{3^{N+1}} > \frac{1}{3^{N-M_*+1}}. \tag{3.4.4}$$

This contradicts (3.4.3) and hence we verify the following claim (I):

(I) If $Q_{w(i)} \subseteq A$, then $Q_{w(i)} \cap \partial R_L = \emptyset$ or $Q_{w(i)} \cap \partial R_R = \emptyset$.

Next we prove that there exists $j \geq 1$ such that $Q_{w(j)} \backslash A \neq \emptyset$. Otherwise, $Q_{w(i)} \subseteq A$ for any $i = 1, \ldots, J + 1$. Let $A_L = [0, a] \times (c, d)$ and let $A_R = [b, 1] \times (c, d)$. Define $I_L = \{i \, | \, i = 1, \ldots, J + 1, Q_{w(i)} \cap A_L \neq \emptyset\}$ and $I_R = \{i \, | \, i = 1, \ldots, J + 1, Q_{w(i)} \cap A_R \neq \emptyset\}$. Since $K_{(w(i))} \subseteq X \cap A \subseteq A_L \cup A_R$, it follows that $\{1, \ldots, J + 1\} = I_L \cup I_R$. Moreover, the claim (I) implies $I_L \cap I_R = \emptyset$. Hence $I_L = \{i \, | \, i = 1, \ldots, J + 1, K_{w(i)} \subseteq A_L\}$ and $I_R = \{i \, | \, i = 1, \ldots, J + 1, K_{w(i)} \subseteq A_R\}$. This contradicts the fact that $(w(1), \ldots, w(J + 1))$ is a chain of K between x and y. Thus there exists $j \geq 1$ such that $Q_{w(j)} \backslash A \neq \emptyset$. Define $i_* = \min\{i \, | \, i = 1, \ldots, J + 1, Q_{w(i)} \backslash A \neq \emptyset\}$. Without loss of generality, we may assume that $Q_{w(i_*)} \cap [0, 1] \times \{d\} \neq \emptyset$. Set

$$\partial R_L^T = \{a\} \times \left[\frac{c + d}{2}, d - \frac{1}{3^{|w(i_*)|}}\right].$$

Shifting $Q_{w(i)}$'s for $i = 1, \ldots, i_* - 1$ horizontally towards ∂R_L, we obtain a covering of ∂R_L^T. Note that the length of ∂R_L^T is $|d - c|/2 - 1/3^{|w(i_*)|}$ and

$$\frac{|d - c|}{2} - \frac{1}{3^{|w(i_*)|}} \geq \frac{\kappa|b - a|}{2} - \frac{1}{3^{N-M_*}} = \frac{\kappa}{2}d(x, y) - \frac{1}{3^{N-M_*}} \geq \frac{\kappa}{2}\frac{1}{3^{N+1}} - \frac{1}{3^{N-M_*}}.$$

On the other hand, the lengths of the sides of $Q_{w(i)}$'s are no less that $1/3^{N-M_*}$ by (3.4.3). Hence

$$i_* - 1 \geq 3^{N-M_*}\left(\frac{\kappa}{2}\frac{1}{3^{N+1}} - \frac{1}{3^{N-M_*}}\right) \geq \frac{\kappa}{2}\frac{1}{3^{M_*+1}} - 1.$$

Since $J + 1 \geq i_*$, it follows that

$$2(J+1)3^{M_*+1} \geq \kappa.$$

This contradicts the fact that κ can be arbitrarily large. Hence we conclude that (SQ4) holds. $\square$

In the followings, we give four examples. The first one has infinite connected components but still the restriction of the Euclidean metric is adapted.

Example 3.4.7 (Figure 3.2) Let X be the self-similar set associated with the contractions $\{F_1, F_3, F_4, F_5, F_7, F_8\}$, i.e. X is the unique nonempty compact set which satisfies

$$X = \bigcup_{i \in S} F_j(X),$$

where $S = \{1, 3, 4, 5, 7, 8\}$. Then $X = C_3 \times [0, 1]$, where C_3 is the ternary Cantor set. Define $(T)_m = S^m$ and $T = \cup_{m \geq 1}(T)_m$. If $K_w = F_w(X)$ for any $w \in T$, then

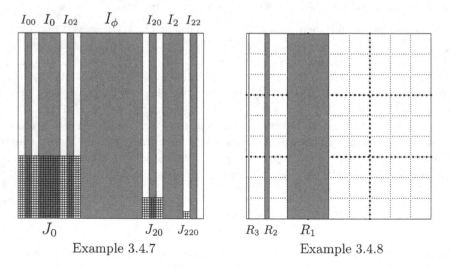

Example 3.4.7   Example 3.4.8

Fig. 3.2 Examples 3.4.7 and 3.4.8

K is a partition of X parametrized by $(T, \mathcal{A}|_T, \phi)$. Define

$$I_\phi = \left[\frac{1}{3}, \frac{2}{3}\right] \times [0, 1] \quad \text{and} \quad I_{i_1,\dots,i_n} = \left[\sum_{k=1}^{n} \frac{i_k}{3^k} + \frac{1}{3^{n+1}}, \sum_{k=1}^{n} \frac{i_k}{3^k} + \frac{2}{3^{n+1}}\right] \times [0, 1]$$

for any $n \geq 1$ and $i_1, \dots, i_n \in \{0, 2\}$. Then

$$\{R_j\}_{j\geq 1} = \{I_\phi, I_{i_1,\dots,i_n} \mid n \geq 1, i_1, \dots, i_n \in \{0, 2\}\}.$$

Set $J_{i_1,\dots,i_n} = [\sum_{k=1}^{n} \frac{i_k}{3^k}, \sum_{k=1}^{n} \frac{i_k}{3^k} + \frac{1}{3^n}] \times [0, \frac{1}{3^n}]$. Then there exists $w \in (T)_n$ such that $J_{i_1,\dots,i_n} = Q_w$, $Q_w \backslash \text{int}(I_{i_1,\dots,i_n})$ has two connected components and $\kappa(Q_w \cap I_{i_1,\dots,i_n}) = 3$. Therefore, $\{R_j\}_{j\geq 1} \subseteq \mathcal{R}_3^1$ and hence d is adapted to $g|_T$.

The second example is the case where the restriction of the Euclidean metric is not adapted.

Example 3.4.8 (Figure 3.2) Set $x_j = \frac{1}{3^j} - \frac{1}{3^{2j}}$, $y_j = \frac{1}{3^j} + \frac{1}{3^{2j}}$ and $R_j = [x_j, y_j] \times [0, 1]$ for any $j \geq 1$. Define $X = Q \backslash (\cup_{j\geq 1} \text{int}(R_j))$. Let $T = \{w \mid w \in W_*, \text{int}(Q_w) \cap X \neq \emptyset\}$ and let $K_w = X \cap Q_w$ for any $w \in T$. Then $K : T \to \mathcal{C}(X)$ is a partition of X parametrized by $(T, \mathcal{A}|_{T\times T}, \phi)$ by Proposition 3.4.1. In this case, we easily see the following facts:

- $\kappa(R_j) = 3^{2j}/2$ for any $j \geq 1$,
- If $w \in \cup_{m\geq j}(T)_m$, then $Q_w \backslash \text{int}(R_j)$ is a rectangle,
- Set $(1)^n = \underbrace{1 \cdots 1}_{n \text{ times}} \in (T)_n$. Then $Q_{(1)^{j-1}} \backslash \text{int}(R_j)$ has two connected components and $\kappa(Q_{(1)^{j-1}} \cap R_j) = 3^{j+1}/2$.

These facts yield that $R_j \notin \mathcal{R}_{3j}^0 \cup \mathcal{R}_{3j}^1$ for sufficiently large j. By Theorem 3.4.4, d is not adapted to $g|_T$. In fact, $D_M^g((x_j, 0), (y_j, 0)) = 3^{-(j-1)}$ for any $j \geq 1$ while $d((x_j, 0), (y_i, 0)) = 2 \cdot 3^{-2j}$. Hence the ratio between $D_M^g(\cdot, \cdot)$ and $d(\cdot, \cdot)$ is not bounded for any $M \geq 0$.

The third one is the case when the restriction of the Euclidean metric is not 1-adapted but 2-adapted.

Example 3.4.9 (Figure 3.3) Define

$$w_*(j) = (1)^{j-1} 9 (1)^j, \quad R_j = Q_{w_*(j)} \quad \text{and} \quad k_m = \left[\frac{m}{2}\right]$$

for $j \in \mathbb{N}$ and $m \in \mathbb{N}$. Note that $(1)^n = \underbrace{1 \dots 1}_{n\text{-times}}$ as is defined in Example 3.4.8. Then it follows that $T = T^{(9)} \backslash \cup_{j\in\mathbb{N}} T_{w_*(j)}^{(9)}$, where $T_w^{(9)} = \{wi_1i_2 \dots \mid i_1, i_2, \dots \in \{1, \dots, 9\}\}$. Let $g(w) = 3^{-|w|}$ for any $w \in T$. Define $w(m) = (1)^{m-1}9$ and $v(m) = (1)^m$. Then $(w(m), (1)^{m-1}8(3)^k, v(m))$ is a chain for $k = 0, 1, \dots, m$. See Fig. 3.3. Therefore $w(m)$ and $v(m)$ are 1-separated in $\Lambda_{3^{-m}}^g$ but not 2-separated in $\Lambda_{3^{-2m}}^g$. This means that the condition $(EV5)_M$ for $M = 1$ does not hold. Therefore, there

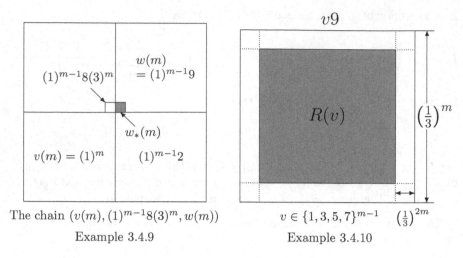

The chain $(v(m), (1)^{m-1}8(3)^m, w(m))$ $v \in \{1, 3, 5, 7\}^{m-1}$ $\left(\frac{1}{3}\right)^{2m}$

Example 3.4.9   Example 3.4.10

Fig. 3.3 Example 3.4.9 and 3.4.10

exists no metric which is 1-adapted to g^α for any $\alpha > 0$. On the other hand, since $\kappa(R_j) = 1$ for any $j \in \mathbb{N}$, the restriction of the Euclidean metric to X, which is denoted by d, is adapted to g. In fact, it is easy to see that d is 2-adapted to g. As a consequence, d is not 1-adapted but 2-adapted to g.

In the fourth example, we do not have thickness while the restriction of the Euclidean metric is adapted.

Example 3.4.10 (Figure 3.3) Define $\Delta Q = (\mathbb{R}^2 \backslash \mathrm{int}(Q)) \cap Q$, which is the topological boundary of Q as a subset of $\mathbb{R}^2$. Let $I_0 = \emptyset$ and let $E = \{1, 3, 5, 7\}$. Define $\{I_n\}_{n \geq 0}$ inductively by $I_{2m-1} = \widehat{I}_{2m-1}$ and $I_{2m} = J_m \cup \widehat{I}_{2m}$ for $m \geq 1$, where

$$J_m = \{v9w | v \in E^{m-1}, w \in W_m, Q_w \cap \Delta Q = \emptyset\}.$$

$\{I_m\}_{m \geq 0}$ satisfies (SQ1), (SQ2) and (SQ3). In fact, if $J_{m,v} = \{v9w | w \in W_m, Q_w \cap \Delta Q = \emptyset\}$ for any $v \in E^{m-1}$, $J_{m,v}$ is a collection of $(3^m - 2)^2$-words in W_{2m}. Set $R(v) = \cup_{u \in J_{m,v}} Q_u$ for any $m \geq 1$ and $v \in E^{m-1}$. See Fig. 3.3. Then $\{R_j\}_{j \geq 1} = \{R(v) | m \geq 1, v \in E^{m-1}\}$. More precisely $R(v) \subseteq Q_{v9}$ and $R(v)$ is a square which has the same center, i.e. the intersection of two diagonals, as Q_{v9} and the length of the sides is $\frac{1}{3^m}(1 - \frac{2}{3^m})$. Note that the length of the sides of Q_{v9} is $\frac{1}{3^m}$. Hence the relative size of $R(v)$ in comparison with Q_{v9} is monotonically increasing and convergent to 1 as $m \to \infty$. The corresponding tree $(T, \mathcal{A}|_T, \phi)$ and the partition $K : T \to \mathcal{C}(X)$ of $X = Q \backslash \cup_{j \geq 1} \mathrm{int}(R_j)$ have the following properties:

Let d be the restriction of the Euclidean metric to X. Then

(a) d is adapted to $g|_T$.
(b) $g|_T$ is exponential and uniformly finite.

(c) Let μ_* be the restriction of the Lebesgue measure on X. Then μ_* has the volume doubling property with respect to d.

(d) μ_* is not gentle with respect to $g|_T$.

(e) μ_* is not super-exponential.

(f) $g|_T$ is not thick.

In the rest, we present proofs of the above claims.

(a) Since $\kappa(R_m) = 1$ for any $m \geq 1$, we see that $\{R_m\}_{m\geq 1} \subseteq \mathcal{R}_1^0$. Hence Theorem 3.4.4 shows that d is adapted to $g|_T$. In fact, d is 1-adapted to $g|_T$ in this case.

(b) This is included in the statement of Proposition 3.4.1-(3).

(c) If $v \in \Lambda_s^{g|T}$ and $Q_v = K_v$, then $\mu_*(K_v) = 9^{-|v|}$ and hence $\mu_*(K_u) \leq 9^{-|u|} = 9^{-|v|+1} \leq 9\mu_*(K_v)$ for any $u \in \Lambda_{3s}^{g|T}$. Therefore, $v \in \Theta(s, 3, k, 9)$ for any $k \geq 1$. On the other hand, for any $w \in T$, there exists $v \in \Lambda_{s,1}^{g|T}(w)$ such that $K_v = Q_v$. Therefore, we see that $\Lambda_{s,1}^{g|T}(w) \cap \Theta(s, 3, 3, 9) \neq \emptyset$. By Lemma 3.3.8, we have (c).

(d) and (e) Set $w(m) = (1)^{m-1}9$. Then $K_{w(m)} = Q_{w(m)} \backslash \mathrm{int}(R_m)$, where $R_m = \cup_{w \in J_m} Q_w$. Then $\mu_*(K_{w(m)}) = 4(3^m - 1)3^{-4m}$. On the other hand, if $v(m) = (1)^{m-1}8$, then $\mu_*(K_{v(m)}) = 3^{-2m}$. Since $K_{w(m)} \cap K_{v(m)} \neq \emptyset$, μ_* is not gentle with respect to $g|_T$. Moreover, since $K_{\pi(w(m))}$ contains $Q_{v(m)}$, we have $\mu_*(K_{\pi(w(m))}) \geq 3^{-2m}$. This implies that μ_* is not super-exponential.

(f) To clarify the notation, we use $B(x, r) = \{y | y \in Q, |x - y| < r\}$ and $B_*(x, r) = B(x, r) \cap X$. This means that $B_*(x, r)$ is the ball of radius r with respect to the metric d on X. Assume that $g|_T$ is thick. Note that d is adapted to $g|_T$. Since K is minimal, Proposition 3.2.2 implies that $K_{w(m)} \supseteq B_*(x, c3^{-m})$ for some $x \in K_{w(m)}$, where c is independent of m and x. However, for any $x \in K_{w(m)}$, there exists $y \in X \backslash K_{w(m)}$ such that $|x - y| \leq 2 \cdot 3^{-2m}$. This contradiction shows that $g|_T$ is not thick.

3.5 Gentleness and Exponentiality

In this section, we show that the gentleness "$\underset{\mathrm{GE}}{\sim}$" is an equivalence relation among exponential weight functions. Moreover, the thickness of the interior, tightness, the uniformly finiteness and the existence of visual metric will be proven to be invariant under the gentle equivalence.

As in the Sect. 3.3, $(T, \mathcal{A}, \phi)$ is a locally finite tree with a reference point ϕ, $(X, \mathcal{O})$ is a compact metrizable topological space with no isolated point and $K : T \rightarrow \mathcal{C}(X, \mathcal{O})$ is a partition of X parametrized by $(T, \mathcal{A}, \phi)$.

Definition 3.5.1 Define $\mathcal{G}_e(T)$ as the collection of exponential weight functions.

Theorem 3.5.2 *The relation $\underset{\mathrm{GE}}{\sim}$ is an equivalence relation on $\mathcal{G}_e(T)$.*

Several steps of preparation are required to prove the above theorem.

Definition 3.5.3

(1) Let $A \subseteq T$. For $m \geq 0$, we define $S^m(A) \subseteq T$ as

$$S^m(A) = \bigcup_{w \in A} ((T)_{m+|w|} \cap T_w).$$

(2) Let $g : T \to (0, 1]$ be a weight function. For any $w \in T$, define

$$N_g(w) = \min\{n | n \geq 0, \pi^n(w) \in \Lambda^g_{g(w)}\}$$

and $\pi^*_g(w) = \pi^{N_g(w)}(w)$.

(3) $(u, v) \in T \times T$ is called an ordered pair if and only if $u \in T_v$ or $v \in T_u$. Define $|u, v| = ||u| - |v||$ for an ordered pair (u, v).

Note that if $g(w) < 1$, then we have

$$N_g(w) = \min\{n | n \geq 0, g(\pi^{n+1}(w)) > g(w)\}.$$

Therefore, if $g(\pi(w)) > g(w)$ for any $w \in T$, then $N_g(w) = 0$ and $\pi^*_g(w) = w$ for any $w \in T$.

The following lemma is immediate from the definitions.

Lemma 3.5.4 *Let $g : T \to (0, 1]$ be a super-exponential weight function, i.e. there exists $\gamma \in (0, 1)$ such that $g(w) \geq \gamma g(\pi(w))$ for any $w \in T$. If (u, v) is an ordered pair, then $g(u) \leq \gamma^{-|u,v|} g(v)$.*

Lemma 3.5.5 *Let $g : T \to (0, 1]$ be a weight function. If g is sub-exponential, then $\sup_{w \in T} N_g(w) < +\infty$.*

Proof Since g is sub-exponential, there exist $c \in (0, 1)$ and $m \geq 0$ such that $cg(w) \geq g(u)$ if $w \in T$, $u \in T_w$ and $|u, v| \geq m$. This immediately implies that $N_g(w) \leq m$.   □

Lemma 3.5.6 *Assume that $g, h \in \mathcal{G}_e(T)$ and h is gentle with respect to g. Then there exist M and N such that if $s \in (0, 1]$, $w \in \Lambda^h_s$ and $u \in S^M(\Lambda^g_{g(w),1}(\pi^*_g(w)))$, then one can choose $n(u) \in [0, N]$ so that $\pi^{n(u)}(u) \in \Lambda^h_s$. Moreover, define $\eta^{g,h}_{s,w} : S^M(\Lambda^g_{g(w),1}(\pi^*_g(w))) \to \Lambda^h_s$ by $\eta^{g,h}_{s,w}(u) = \pi^{n(u)}(u)$. Then $\Lambda^h_{s,1}(w) \subseteq \eta^{g,h}_{s,w}(S^M(\Lambda^g_{g(w),1}(\pi^*_g(w))))$. In particular, for any $s \in (0, 1]$, $w \in \Lambda^h_s$ and $v \in \Lambda^h_{s,1}(w)$, there exists $u \in \Lambda^g_{g(w),1}(\pi^*_g(w))$ such that (u, v) is an ordered pair and $|u, v| \leq \max\{M, N\}$.*

Proof Since h is sub-exponential, there exist $c_1 \in (0, 1)$ and $m \geq 0$ such that $c_1 h(w) \geq h(u)$ for any $w \in T$ and $u \in S^m(w)$. Let $w \in \Lambda^h_s$ and let $w' = \pi^*_g(w)$. Set $t = g(w)$. Let $v \in \Lambda^g_{t,1}(w')$. As h is gentle with respect to g, there exists $c \geq 1$

such that

$$h(w')/c \leq h(v) \leq ch(w'),$$

where c is independent of s, w and v. By Lemma 3.5.5 and the fact that h is super-exponential, there exists $c' \geq 1$ such that

$$h(w)/c \leq h(v) \leq c'h(w)$$

for any s, w and v. Using this, h being sub-exponential and Proposition 3.1.16, we see that there exist $c'' > 0$ and M, which are independent of s and w, such that $c''s \leq h(u) \leq s$ for any $u \in S^M(\Lambda^g_{t,1}(w'))$. Choose k so that $c''(c_1)^{-k} > 1$. Then $h(\pi^{km}(u)) \geq (c_1)^{-k}h(u) \geq c''(c_1)^{-k}s > s$. Set $N = km - 1$. Then, for any $u \in S^M(\Lambda^g_{t,1}(w'))$, there exists $n(u)$ such that $n(u) \leq N$ and $\pi^{n(u)}(u) \in \Lambda^h_s$. Now for any $\rho \in \Lambda^h_{s,1}(w)$, there exists $v \in \Lambda^g_{t,1}(w')$ such that (ρ, v) is an ordered pair. Since $\pi^{n(u)}(u) = \rho$ for any $u \in S^M(v)$, it follows that $\eta^{g,h}_{s,w}(S^M(\Lambda^g_{g(w),1}(\pi^*_g(w)))) \supseteq \Lambda^h_{s,1}(w)$. The rest is straightforward. $\qquad\square$

Finally we are ready to give a proof of Theorem 3.5.2.

Proof of Theorem 3.5.2 Let $g, h, \xi \in \mathcal{G}_e(T)$. Then there exists $\gamma \in (0, 1)$ such that $g(w) \geq \gamma g(\pi(w))$, $h(w) \geq \gamma h(\pi(w))$ and $\xi(w) \geq \gamma \xi(\pi(w))$ for any $w \in T$.

First we show $g \underset{GE}{\sim} g$. By Proposition 3.1.16, there exists $c \in (0, 1)$ such that if $w \in \Lambda^g_s$, then $cg(w) \leq s \leq g(w)$. As a consequence, if $w, v \in \Lambda^g_s$, then $g(w) \leq s/c \leq g(v)/c$. Thus $g \underset{GE}{\sim} g$.

Next assume $h \underset{GE}{\sim} g$. Suppose that $w, v \in \Lambda^h_s$ and $K_w \cap K_v \neq \emptyset$. Since $v \in \Lambda^h_{s,1}(w)$, Lemma 3.5.6 implies that there exists $u \in \Lambda^g_{g(w),1}(\pi^*_g(w))$ such that (u, v) is an ordered pair and $|u, v| \leq L$, where $L = \max\{M, N\}$. By Lemma 3.5.4, $g(v) \geq \gamma^L g(u) \geq \gamma^L g(w)$. Hence $g \underset{GE}{\sim} h$.

Finally assume that $h \underset{GE}{\sim} g$ and $\xi \underset{GE}{\sim} h$. Suppose that $w, v \in \Lambda^\xi_s$ and $K_w \cap K_v \neq \emptyset$. Since $v \in \Lambda^\xi_{s,1}(w)$, Lemma 3.5.6 implies that there exists $u \in \Lambda^h_{h(w),1}(\pi^*_h(w))$ such that (u, v) is an ordered pair and $|u, v| \leq L$. By Lemma 3.5.4, it follows that $g(v) \geq \gamma^L g(u)$. Set $s' = h(w)$ and $w' = \pi^*_h(w)$. Note that $w' \in \Lambda^h_{s'}$ and $u \in \Lambda^h_{s',1}(w')$. Again by Lemma 3.5.6, there exists $a \in \Lambda^g_{g(w'),1}(\pi^*_g(w'))$ such that (u, a) is an ordered pair and $|a, u| \leq L$. Lemma 3.5.4 shows that $g(u) \geq \gamma^L g(a) \geq \gamma^L g(\pi^*_h(w))$. Note that $|\pi^*_h(w)| \leq |w|$ and hence $g(\pi^*_h(w)) \geq g(w)$. Combining these, we obtain $g(v) \geq \gamma^{2L} g(\pi^*_h(w)) \geq \gamma^{2L} g(w)$. Hence $\xi \underset{GE}{\sim} g$. Consequently we verify $g \underset{GE}{\sim} \xi$ by the above arguments. $\qquad\square$

Next, we show the invariance of thickness, tightness and uniform finiteness under the equivalence relation $\underset{GE}{\sim}$.

Theorem 3.5.7 *Let $g, h \in \mathcal{G}_e(T)$. Suppose $g \underset{\text{GE}}{\sim} h$.*

(1) *Suppose that $\sup_{w \in T} \#(S(w)) < +\infty$. If g is uniformly finite then so is h.*
(2) *If g is thick, then so is h.*
(3) *If g is tight, then so is h.*

We need the next lemma to prove Theorem 3.5.7.

Lemma 3.5.8 *Let $g, h \in \mathcal{G}_e(T)$. Assume that g is gentle with respect to h. Then for any $\alpha \in (0, 1]$ and $M \geq 0$, there exists $\gamma \in (0, 1)$ such that*

$$U_M^g(x, \alpha g(w)) \supseteq U_M^h(x, \gamma h(w))$$

for any $w \in T$ and $x \in K_w$.

Proof Since g and h are exponential, there exist $c_1, c_2 \in (0, 1)$ and $m \geq 1$ such that $h(w) \geq c_2 h(\pi(w))$, $g(w) \geq c_2 g(\pi(w))$, $h(v) \leq c_1 h(w)$ and $g(v) \leq c_1 g(w)$ for any $w \in T$ and $v \in S^m(w)$. Moreover, since g is gentle with respect to h, there exists $c > 1$ such that $g(w) \leq cg(u)$ whenever $w, u \in \Lambda_s^h$ and $K_w \cap K_v \neq \emptyset$. Note that $N_g(w) \leq m$ and $N_h(w) \leq m$ for any $w \in T$.

Let $w \in T$ and let $x \in K_w$. Assume that $\gamma < (c_2)^{lm}$. Let $v \in \Lambda_{\gamma h(w),0}^h(x)$. Then $h(\pi(v)) > \gamma h(w) \geq h(v)$. There exists $k \geq 0$ such that $\pi^k(v) \in \Lambda_{h(w)}^h$. Then $h(\pi^{k+1}(v)) > h(w) \geq h(\pi^k(v))$. Thus we have

$$\gamma h(\pi^{k+1}(v)) \geq h(v).$$

Therefore, it follows that $k + 1 \geq lm$. Let $w_* = \pi^{N_h(w)}(w)$. Then we see that $x \in K_{\pi^{k+1}(v)} \cap K_{w_*}$. Therefore, $c^{-1} g(w_*) \leq g(\pi^{k+1}(v)) \leq cg(w_*)$. Since $k + 1 \geq lm$ and $N_h(w) \leq m$, it follows that

$$g(v) \leq (c_1)^l g(\pi^{k+1}(v)) \leq c(c_1)^l g(w_*) \leq c(c_1)^l (c_2)^{-m} g(w).$$

Now suppose that $(w(1), \ldots, w(M + 1))$ is a chain in $\Lambda_{\gamma h(w)}^h$ with $w(1) \in \Lambda_{\gamma h(w),0}^h(x)$. Using the above arguments, we obtain

$$g(w(i)) \leq c^{i-1} g(w(1)) \leq c^i (c_1)^l (c_2)^{-m} g(w) \leq c^{M+1} (c_1)^l (c_2)^{-m} g(w)$$

for any $i = 1, \ldots, M + 1$. Choosing l so that $c^{M+1}(c_1)^l (c_2)^{-m} < \alpha$, we see that $U_M^h(x, \gamma h(w)) \subseteq U_M^g(x, \alpha g(w))$. $\qquad \square$

Proof of Theorem 3.5.7 (1) Set $L = \sup_{w \in T} \#(S(w))$. By Lemma 3.5.6, it follows that $\#(\Lambda_{s,1}^h(w)) \leq L^M \#(\Lambda_{g(w),1}^g(\pi_g^*(w)))$. This suffices to the desired conclusion.

(2) By the thickness of g and Proposition 3.2.1, for any $M \geq 0$, there exists $\beta > 0$ such that, for any $w \in T$,

$$K_w \supseteq U_M^g(x, \beta g(\pi(w)))$$

for some $x \in K_w$. By Lemma 3.5.8, there exists $\gamma \in (0, 1)$ such that

$$U_M^g(x, \beta g(\pi(w))) \supseteq U_M^h(x, \gamma h(\pi(w)))$$

for any $w \in T$. Thus making use of Proposition 3.2.1 again, we see that h is thick.

(3) By the remark after the proof of Proposition 3.1.12, for any $M \geq 0$, there exists $\alpha > 0$ such that, for any $w \in T$, $K_w \backslash U_M^g(x, \alpha g(w)) \neq \emptyset$ for some $x \in K_w$. By Lemma 3.5.8, there exists $\gamma \in (0, 1)$ such that $U_M^g(x, \alpha g(w)) \supseteq U_M^h(x, \gamma h(w))$ for any $w \in T$ and $x \in K_w$. Hence

$$\sup_{x, y \in K_w} \delta_M^h(x, y) \geq \gamma h(w)$$

for any $w \in T$. Thus we have shown that h is tight. $\square$

Finally, the existence of visual metric is also invariant under $\underset{\text{GE}}{\sim}$ as follows.

Theorem 3.5.9 *Assume that the partition $K : T \to \mathcal{C}(X, \mathcal{O})$ is minimal. Let $g, h \in \mathcal{G}_e(T)$. Assume that $g \underset{\text{GE}}{\sim} h$. Then g is hyperbolic if and only if h is hyperbolic.*

Proof Since g and h are exponential, there exist $\lambda \in (0, 1)$ and $m \geq 1$ such that

$$g(w') \leq \lambda g(w) \leq g(w'') \quad \text{and} \quad h(w') \leq \lambda h(w) \leq h(w'')$$

if $w \in T$, $w', w'' \in T_w$, $|w'| - |w| \geq m$ and $|w''| - |w| = 1$. Moreover, since $g \underset{\text{GE}}{\sim} h$, there exists $\eta > 1$ such that if $w, v \in \Lambda_s^g$ and $K_w \cap K_v \neq \emptyset$, then $h(w) \leq \eta h(v)$ and if $w, v \in \Lambda_s^h$ and $K_w \cap K_v \neq \emptyset$, then $g(w) \leq \eta g(v)$.

Now assume that g is hyperbolic. Then by Theorem 2.4.12 and 2.5.12, g satisfies (EV5)$_M$ for some $M \geq 1$. Fix $k = k(L) \in \mathbb{N}$ satisfying $\eta^L \lambda^k < 1$. Let $w, v \in \Lambda_s^h$ and assume that (w, v) is L-separated in Λ_s^h. Set $t = g(v)$. Suppose that (w, v) is not L-separated in $\Lambda_{\lambda^{km}t}^g$. Then there exists a chain $(w_*(1), \ldots, w_*(L-1))$ in $\Lambda_{\lambda^{km}t}^g$ such that $(w, w_*(1), \ldots, w_*(L-1), v)$ is a chain. Choose $v_* \in \Lambda_{\lambda^{km}t}^g \cap T_v$ so that $K_{w_*(L-1)} \cap K_{v_*} \neq \emptyset$. Since $g(v_*) \leq \lambda^{km}t = \lambda^{km}g(v)$, it follows that $|v_*| - |v| \geq km$. Then we have

$$h(w_*(i)) \leq \eta^L h(v_*) \leq \eta^L \lambda^k h(v) < h(v).$$

Hence there exists a chain $(w(1), \ldots, w(L-1))$ in Λ_s^h such that $w_*(i) \in T_{w(i)}$ for any $i = 1, \ldots, L-1$. This implies that (w, v) is not L-separated in Λ_s^h. This contradiction implies that (w, v) is L-separated in $\Lambda^g_{\lambda k m_t}$. Thus we have shown the following claim.

Claim 1 Let $w, v \in \Lambda^h_s$. If (w, v) is L-separated in Λ^h_s, then it is L-separated in $\Lambda^g_{\lambda k(L)m} g(v)$.

Exchanging g and h and using the same arguments, we obtain the following claim as well.

Claim 2 Let $w, v \in \Lambda^g_s$. If (w, v) is L-separated in Λ^g_s, then it is L-separated in $\Lambda^g_{\lambda k(L)m} h(v)$.

Now, assume that $w, v \in \Lambda^h_s$ and (w, v) is M-separated in Λ^h_s. By Claim 1, we see that (w, v) is M-separated in $\Lambda^g_{\lambda k(M)m} g(v)$. Since $(EV5)_M$ holds for g, there exists $\tau \in (0, 1)$ such that if $u, u' \in \Lambda^g_{s'}$ and (u, u') is M-separated in $\Lambda^g_{s'}$, then it is $(M + 1)$-separated in $\Lambda^g_{\tau s'}$. This shows that (w, v) is $(M + 1)$-separated in $\Lambda^g_{\tau \lambda k(M)m} g(v)$. Set $t_* = \tau \lambda^{k(M)m} g(v)$. If $w' \in \Lambda^g_{t_*} \cap T_w$ and $v' \in \Lambda^g_{t_*} \cap T_v$, then (w', v') is $(M + 1)$-separated in $\Lambda^g_{t_*}$. So applying Claim 2, we see that (w', v') is $(M + 1)$-separated in $\Lambda^h_{\lambda k(M+1)m} h(v')$. Choose $v' \in \Lambda^g_{t_*} \cap T_v$ so that $h(v') = \min\{g(v'') | v'' \in \Lambda^g_{t_*} \cap T_v\}$, then it follows that (w, v) is $(M + 1)$-separated in $\Lambda^h_{\lambda k(M+1)m} h(v')$.

Note that $g(w) \geq \lambda g(\pi(w)) > \lambda s$ if $w \in \Lambda^g_s$. Choose n_* so that $\lambda^{n_*-1} < \tau$. Suppose $|v'| - |v| \geq (k(M)m + n_*)m$. Set $k = k(M)$ and $k' = k(M + 1)$. Then

$$\lambda^{km+n_*} g(v) < \lambda \tau \lambda^{km} g(v) = \lambda t_* \leq g(v') \leq \lambda^{km+n_*} g(v).$$

This contradiction yields that $|v'| - |v| < (km + n_*)m$. Therefore, $h(v') \geq \lambda^{(km+n_*)m} h(v) \geq \lambda^{(km+n_*)m+1} s$. Thus $\lambda^{k'm} h(v') \geq \lambda^{(km+n_*+k')m+1} s$. Set $\tau_* = \lambda^{(km+n_*+k')m+1}$. Then (w, v) is $(M + 1)$-separated in $\Lambda^h_{\tau_* s}$. Thus we have shown that $(EV5)_M$ is satisfied for h. Using Theorems 2.4.12 and 2.5.12, we see that h is hyperbolic.

3.6 Quasisymmetry

In this section, we are going to identify the equivalence relation, gentleness "$\underset{GE}{\sim}$" with the quasisymmetry "$\underset{QS}{\sim}$" among the metrics under certain conditions. As in the last section, $(T, \mathcal{A}, \phi)$ is a locally finite tree with a reference point ϕ, $(X, \mathcal{O})$ is a compact metrizable topological space with no isolated point and $K : T \to \mathcal{C}(X, \mathcal{O})$ is a partition of X parametrized by $(T, \mathcal{A}, \phi)$ throughout this section.

Definition 3.6.1 (Quasisymmetry) A metric $\rho \in \mathcal{D}(X, \mathcal{O})$ is said to be quasisymmetric to a metric $d \in \mathcal{D}(X, \mathcal{O})$ if and only if there exists a homeomorphism h from

$[0, +\infty)$ to itself such that $h(0) = 0$ and, for any $t > 0$, $\rho(x, z) < h(t)\rho(x, y)$ whenever $d(x, z) < td(x, y)$. We write $\rho \underset{QS}{\sim} d$ if ρ is quasisymmetric to d.

It is known that $\underset{QS}{\sim}$ is an equivalence relation on $\mathcal{D}(X, \mathcal{O})$.

Definition 3.6.2 Let $d \in \mathcal{D}(X, \mathcal{O})$. We say that d is (super-, sub-)exponential if and only if g_d is (super-, sub-)exponential.

Under the uniformly perfectness of a metric space defined below, we can utilize a useful equivalent condition for quasisymmetry obtained in [21]. See the details in the proof of Theorem 3.6.4.

Definition 3.6.3 A metric space (X, d) is called uniformly perfect if and only if there exists $\epsilon > 0$ such that $B_d(x, (1 + \epsilon)r) \backslash B_d(x, r) \neq \emptyset$ unless $B_d(x, r) = X$.

Lemma 3.6.4 *Let $d \in \mathcal{D}(X, \mathcal{O})$. If d is super-exponential, then (X, d) is uniformly perfect.*

Proof Write $d_w = g_d(w)$ for any $w \in T$. Since d is super-exponential, there exists $c_2 \in (0, 1)$ such that $d_w \geq c_2 d_{\pi(w)}$ for any $w \in T$. Therefore, $s \geq d_w > c_2 s$ if $w \in \Lambda_s^d$. For any $x \in X$ and $r \in (0, 1]$, choose $w \in \Lambda_{r/2,0}^d(x)$. Then $d(x, y) \leq d_w \leq r/2$ for any $y \in K_w$. This shows $K_w \subseteq B_d(x, r)$. Since $\text{diam}(B_d(x, c_2 r/4), d) \leq c_2 r/2 < d_w$, it follows that $K_w \backslash B_d(x, c_2 r/2) \neq \emptyset$. Therefore $B_d(x, r) \backslash B_d(x, c_2 r/2) \neq \emptyset$. This shows that (X, d) is uniformly perfect. $\square$

Definition 3.6.5 Define

$$\mathcal{D}_{A,e}(X, \mathcal{O}) = \{d \,|\, d \in \mathcal{D}(X, \mathcal{O}), d \text{ is adapted and exponential.}\}$$

The next theorem is the main result of this section.

Theorem 3.6.6 *Let $d \in \mathcal{D}_{A,e}(X, \mathcal{O})$ and let $\rho \in \mathcal{D}(X, \mathcal{O})$. Then $d \underset{QS}{\sim} \rho$ if and only if $\rho \in \mathcal{D}_{A,e}(X, \mathcal{O})$ and $d \underset{GE}{\sim} \rho$. Moreover, if d is M-adapted and $d \underset{QS}{\sim} \rho$, then ρ is M-adapted as well.*

Remark In the case of natural partitions of self-similar sets introduced in Example 2.2.7, the above theorem has been obtained in [22].

The following corollary is straightforward from the above theorem.

Corollary 3.6.7 *Let $d, \rho \in \mathcal{D}_{A,e}(X, \mathcal{O})$. Then $d \underset{QS}{\sim} \rho$ if and only if $d \underset{GE}{\sim} g$.*

The rest of this section is devoted to a proof of the above theorem.

Proof of Theorem 3.6.6: Part 1 Assume that $d, \rho \in \mathcal{D}_{A,e}(X, \mathcal{O})$. We are going to show that if $d \underset{GE}{\sim} \rho$, then $d \underset{QS}{\sim} \rho$. By Lemma 3.6.4, both (X, d) and (X, ρ) are uniformly perfect. By [21, Theorems 11.5 and 12.3], $d \underset{QS}{\sim} \rho$ is equivalent to the fact that there exists $\delta \in (0, 1)$ such that

$$B_d(x, r) \supseteq B_\rho(x, \delta\overline{\rho}_d(x, r)) \quad \text{and} \quad B_\rho(x, r) \supseteq B_d(x, \delta\overline{d}_\rho(x, r)) \tag{3.6.1}$$

and

$$\overline{\rho}_d(x, r/2) \geq \delta\overline{\rho}_d(x, r) \quad \text{and} \quad \overline{d}_\rho(x, r/2) \geq \delta\overline{d}_\rho(x, r) \tag{3.6.2}$$

for any $x \in X$ and $r > 0$, where $\overline{\rho}_d(x, r) = \sup_{y \in B_d(x,r)} \rho(x, y)$ and $\overline{d}_\rho(x, r) = \sup_{y \in B_\rho(x,r)} d(x, y)$. Our aim is to show (3.6.1) and (3.6.2). Since d and ρ are adapted, there exist $\beta \in (0, 1)$, $\gamma > 1$ and $M \geq 1$ such that

$$U_M^d(x, \beta r) \subseteq B_d(x, r) \subseteq U_M^d(x, \gamma r) \quad \text{and} \quad U_M^\rho(x, \beta r) \subseteq B_\rho(x, r) \subseteq U_M^\rho(x, \gamma r)$$

for any $x \in X$ and $r \in (0, 1]$. By Lemma 3.5.8, there exists $\alpha \in (0, 1)$ such that $U_M^\rho(x, \rho_w) \supseteq U_M^d(x, \alpha d_w)$ and $U_M^d(x, d_w) \supseteq U_M^\rho(x, \alpha\rho_w)$ for any $w \in T$ and $x \in K_w$. If $w \in \Lambda_{\gamma r/\alpha, 0}^d(x)$, then

$$B_d(x, r) \subseteq U_M^d(x, \gamma r) \subseteq U_M^d(x, \alpha d_w) \subseteq U_M^\rho(x, \rho_w). \tag{3.6.3}$$

Hence for any $y \in B_d(x, r)$, there exists $(w(1), \ldots, w(k)) \in \mathcal{CH}_K(x, y)$ such that $k \leq M + 1$ and $w(i) \in \Lambda_{\rho_w}^\rho$. Since $\rho(x, y) \leq \sum_{i=1}^k \rho_{w(i)} \leq (M+1)\rho_w$, we have

$$\overline{\rho}_d(x, r) \leq (M+1)\rho_w.$$

Let $w \in \Lambda_{\gamma r/\alpha, 0}^d(x)$ as above. Since $\beta/2 < 1 < \gamma/\alpha$, there exists $v \in T_w$ such that $v \in \Lambda_{\beta r/2, 0}^d(x)$. Note that $\beta r/2 \geq d_v$. Hence we have

$$B_d\left(x, \frac{r}{2}\right) \supseteq U_M^d\left(x, \frac{\beta r}{2}\right) \supseteq U_M^d(x, d_v) \supseteq U_M^\rho(x, \alpha\rho_v). \tag{3.6.4}$$

Since d is sub-exponential, the fact that $w \in \Lambda_{\gamma r/\alpha, 0}^d(x)$ and $v \in \Lambda_{\beta r/2, 0}^d(x) \cap T_w$ implies that $|v| - |w|$ is uniformly bounded with respect to x, r and w. This and the fact that ρ is super-exponential imply that there exists $c > 0$, which is independent of x, r and w, such that $\rho_v \geq c\rho_w$. Now we see that $\alpha\rho_v \geq \eta\overline{\rho}_d(x, r)$, where $\eta = \alpha c/(M+1)$. Hence by (3.6.4),

$$B_d\left(x, \frac{r}{2}\right) \supseteq U_M^\rho(x, \eta\overline{\rho}_d(x, r)) \supseteq B_\rho\left(x, \frac{\eta}{\gamma}\overline{\rho}_d(x, r)\right). \tag{3.6.5}$$

By the fact that (X, ρ) is uniformly perfect, there exists $c_* \in (0, 1)$ such that $B_\rho(y, t) \backslash B_\rho(y, c_* t) \neq \emptyset$ unless $B_\rho(y, c_* t) = X$. Set $\delta = c_*\eta/\gamma$. Note that $\delta < \eta/\gamma < 1$. Since $\delta\overline{\rho}_d(x, r) \leq \delta\sup_{y \in X} \rho(x, y) < \sup_{y \in X} \rho(x, y)$, there exists $y \in X$ such that $\rho(x, y) > \delta\overline{\rho}_d(x, r)$. Hence $B_\rho(x, \delta\overline{\rho}_d(x, r)) \neq X$.

This shows that $B_\rho(x, \eta \overline{\rho}_d(x, r)/\gamma) \setminus B_\rho(x, \delta \overline{\rho}_d(x, r)) \neq \emptyset$. By (3.6.5), there exists $z \in B_d(x, r/2)$ such that $\rho(x, z) \geq \delta \overline{\rho}_d(x, r)$ and hence $\overline{\rho}_d(x, r/2) \geq \delta \overline{\rho}_d(x, r)$. Furthermore, $B_d(x, r) \supseteq B_\rho(x, \eta \overline{\rho}_d(x, r)/\gamma) \supseteq B_\rho(x, \delta \overline{\rho}_d(x, r))$. Thus we have obtained halves of (3.6.1) and (3.6.2). Exchanging d and ρ, we have the other halves of (3.6.1) and (3.6.2). □

Lemma 3.6.8 *Let* $d \in \mathcal{D}_{A,e}(X, \mathcal{O})$ *and let* $\rho \in \mathcal{D}(X, \mathcal{O})$. *Assume that* $d \underset{QS}{\sim} \rho$. *Then* (3.6.1) *and* (3.6.2) *hold. Moreover, let* $\delta \in (0, 1)$ *be the constant appearing in* (3.6.1) *and* (3.6.2).

(1) *For any* $w \in T$ *and* $x, y \in K_w$,

$$\overline{\rho}_d(x, d_w) \leq \delta^{-1} \overline{\rho}_d(y, d_w).$$

(2) *There exists* $c > 0$ *such that*

$$c \overline{\rho}_d(x, d_w) \leq \rho_w \leq \delta^{-1} \overline{\rho}_d(x, d_w)$$

for any $w \in T$ *and* $x \in K_w$.

Proof Assume $d \underset{QS}{\sim} \rho$. Lemma 3.6.4 implies that (X, d) is uniformly perfect. Since uniformly perfectness is preserved by quasisymmetry, see a proof of this fact in [21, Proposition 12.2], (X, ρ) is uniformly perfect as well. Hence as we have mentioned above, the fact that $d \underset{QS}{\sim} \rho$ implies (3.6.1) and (3.6.2).

(1) Since $B_d(x, d_w) \subseteq B_d(y, 2d_w)$, it follows that $\overline{\rho}_d(x, d_w) \leq \overline{\rho}_d(y, 2d_w)$. Applying (3.6.2), we obtain the desired inequality.

(2) For any $x \in K_w$, $K_w \subseteq B_d(x, 2d_w)$. Hence $\rho_w \leq \overline{\rho}_d(x, 2d_w)$. By (3.6.2), we see that

$$\rho_w \leq \delta^{-1} \overline{\rho}_d(x, d_w).$$

Set $s = d_w/2$ and choose $v \in T_w \cap \Lambda_s^d$. Since d is adapted and tight, there exists $\gamma > 0$, which is independent of w, v and s, such that

$$K_v \setminus B_d(z, \gamma d_v) \neq \emptyset$$

for some $z \in K_v$. By (3.6.1),

$$K_v \setminus B_\rho(z, \delta \overline{\rho}_d(z, \gamma d_v)) \neq \emptyset.$$

Hence $\rho_w \geq \delta \overline{\rho}_d(z, \gamma d_v)$. Since d is super-exponential, there exists $\gamma' > 0$, which is independent of w, v and s, such that $\gamma d_v \geq \gamma' d_w$. Choose $n \geq 1$ so that $2^{n-1}\gamma' \geq 1$. Using (3.6.2) n-times, we have

$$\rho_w \geq \delta \overline{\rho}_d(z, \gamma' d_w) \geq \delta^{n+1} \overline{\rho}_d(z, d_w).$$

By (1), if $c = \delta^{n+2}$, then $\rho_w \geq c\overline{\rho}_d(x, d_w)$.   □

Proof of Theorem 3.6.6: Part 2 Assume that $d \in \mathcal{D}_{A,e}(X, \mathcal{O})$. We show that if $d \underset{\text{QS}}{\sim} \rho$, then $\rho \in \mathcal{D}_{A,e}(X, \mathcal{O})$ and $d \underset{\text{GE}}{\sim} \rho$. As in the proof of Lemma 3.6.8, (3.6.1) and (3.6.2) hold.

Claim 1 ρ is super-exponential.

Proof of Claim 1 Since d is super-exponential, there exists $c' \in (0, 1)$ such that $d_w \geq c' d_{\pi(w)}$ for any $w \in T$. Choose $l \geq 1$ so that $2^l c' \geq 1$. By Lemma 3.6.8-(2) and (3.6.2), if $x \in K_w$, then

$$\rho_w \geq c\overline{\rho}_d(x, d_w) \geq c\delta^l \overline{\rho}_d(z, 2^l d_w) \geq c\delta^l \overline{\rho}_d(x, d_{\pi(w)}) \geq c\delta^{l+1} \rho_{\pi(w)}.$$

Claim 2 ρ is sub-exponential.

Proof of Claim 2 Since d is sub-exponential, there exist $c_1 \in (0, 1)$ and $m \geq 1$ such that

$$d_{v'} \leq c_1 d_w$$

for any $w \in T$ and $v' \in S^m(w)$. Let $w \in T$. If $v \in S^{mj}(w)$ for $j \geq 1$ and $x \in K_v$, then by Lemma 3.6.8-(2)

$$\rho_v \leq \delta^{-1} \overline{\rho}_d(x, d_v) \leq \delta^{-1} \overline{\rho}_d(x, (c_1)^j d_w). \tag{3.6.6}$$

On the other hand, since $d \underset{\text{QS}}{\sim} \rho$, the $(\text{SQS})_d$ condition in [21] is satisfied by [21, Theorem 12.3]. Then by [21, Proposition 11.7], there exist $\lambda \in (0, 1)$ and $c'' > 0$ such that

$$\overline{\rho}_d(x, c_1 s) \leq c'' \lambda \overline{\rho}_d(x, s)$$

for any $x \in X$ and $s \in (0, 1]$. This together with (3.6.6) and Lemma 3.6.8-(2) yields

$$\rho_v \leq \delta^{-1} \overline{\rho}_d(x, (c_1)^j d_w) \leq \delta^{-1} c'' \lambda^j \overline{\rho}_d(x, d_w) \leq \delta^{-1} c'' \lambda^j c^{-1} \rho_w.$$

Choosing j so that $\delta^{-1} c'' \lambda^j c^{-1} < 1$, we see that ρ is sub-exponential.

Claim 3 $d \underset{\text{GE}}{\sim} \rho$.

Proof of Claim 3 Since d is super-exponential, there exists $c_2 \in (0, 1)$ such that

$$s \geq d_w > c_2 s \tag{3.6.7}$$

for any $s \in (0, 1]$ and $w \in \Lambda_s^d$. Let $w, v \in \Lambda_s^d$ with $K_w \cap K_v \neq \emptyset$. Then $d_w \leq d_v/c_2$. Choose $k \geq 1$ so that $2^k c_2 \geq 1$. If $x \in K_w \cap K_v$, then by Lemma 3.6.8-(2) and (3.6.2),

$$\rho_w \leq \delta^{-1}\overline{\rho}_d(x, d_w) \leq \delta^{-1}\overline{\rho}(x, d_v/c_2) \leq \delta^{-(k+1)}\overline{\rho}(x, d_v) \leq c^{-1}\delta^{-(k+1)}\rho_v.$$

Hence $d \underset{\mathrm{GE}}{\sim} \rho$.

Claim 4 ρ is adapted. More precisely, if d is M-adapted, then so is ρ.

Proof of Claim 4 Assume that d is M-adapted. Let $x \in X$ and let $s \in (0, 1]$. Then there exists $\alpha > 1$, which is independent of x and s, such that $U_M^d(x, \alpha s) \supseteq B_d(x, s)$. Let $w \in \Lambda_{s,0}^\rho(x)$. Since ρ is super-exponential, there exists $b \in (0, 1)$, which is independent of w and s, such that $\rho_w \geq bs$. By Lemma 3.5.8, there exists $\gamma \in (0, 1)$ such that $U_M^\rho(x, \rho_w) \supseteq U_M^d(x, \gamma d_w)$ for any $w \in T$ and $x \in K_w$. Choose $p \geq 1$ so that $2^p \gamma/\alpha \geq 1$. Then by Lemma 3.6.8-(2), (3.6.1) and (3.6.2),

$$U_M^\rho(x, s) \supseteq U_M^\rho(x, \rho_w) \supseteq U_M^d(x, \gamma d_w)$$

$$\supseteq B_d\left(x, \frac{\gamma}{\alpha}d_w\right) \supseteq B_\rho\left(x, \delta\overline{\rho}_d\left(x, \frac{\gamma}{\alpha}d_w\right)\right) \supseteq B_\rho(x, \delta^{p+1}\overline{\rho}_d(x, d_w))$$

$$\supseteq B_\rho(x, \delta^{p+2}\rho_w) \supseteq B_\rho(x, \delta^{p+2}bs).$$

On the other hand, let $x \in K$ and let $r \in (0, 1]$. Then for any $y \in U_M^\rho(x, r)$, there exists $(w(1), \ldots, w(M+1)) \in \mathcal{CH}_K(x, y)$ such that $w(i) \in \Lambda_r^\rho$ for any i. It follows that

$$\rho(x, y) \leq \sum_{i=1}^{M+1} \rho_{w(i)} \leq (M+1)r.$$

This shows that $U_M^\rho(x, r) \subseteq B_\rho(x, (M+1)r)$. Thus we have shown that ρ is M-adapted. $\qquad\square$

Chapter 4
Characterization of Ahlfors Regular Conformal Dimension

4.1 Construction of Adapted Metric I

In this section, we present a sufficient condition for the existence of an adapted metric to a given weight function. The sufficient condition obtained in this section will be used to construct an Ahlfors regular metric later.

Let $(T, \mathcal{A}, \phi)$ be a locally finite tree with a reference point ϕ and let $(X, \mathcal{O})$ be a compact metrizable topological space with no isolated point. Moreover let $K : T \to \mathcal{C}(X, \mathcal{O})$ be a minimal partition.

Definition 4.1.1 Let $M \geq 1$. A chain $(w(1), \ldots, w(M+1))$ of K is called a horizontal M-chain of K if and only if $|w(i)| = |w(i+1)|$ and $K_{w(i)} \cap K_{w(i+1)} \neq \emptyset$ for any $i = 1, \ldots, M$. Define

$$\Gamma_M(w, K) = \{u | u \in (T)_{|w|}, \text{ there exists a horizontal } M\text{-chain}$$

$$(w(1), \ldots, w(M+1)) \text{ of } K \text{ such that}$$

$$w(1) = w \text{ and } w(M+1) = u\}$$

for $w \in T$,

$$J^h_{M,n}(K) = \{(w, u) | w, u \in (T)_n, w \neq v, u \in \Gamma_M(w, K)\}$$

for $n \geq 0$,

$$J^h_M(K) = \bigcup_{n \geq 0} J^h_{M,n}(K),$$

$$J^v_M(K) = \{(w, u) | w, u \in T, w \in \pi(\Gamma_M(u, K)) \text{ or } u \in \pi(\Gamma_M(w, K))\},$$

© The Editor(s) (if applicable) and The Author(s), under exclusive license to Springer Nature Switzerland AG 2020
J. Kigami, *Geometry and Analysis of Metric Spaces via Weighted Partitions*, Lecture Notes in Mathematics 2265, https://doi.org/10.1007/978-3-030-54154-5_4

and

$$J_M(K) = J_M^v(K) \cup J_M^h(K).$$

A sequence $(w(1), \dots, w(m)) \in T$ is called an M-jumping path, or an M-jpath for short, (resp. horizontal M-jumping path, or horizontal M-jpath) of K if and only if $(w(i), w(i+1)) \in J_M(K)$ (resp. $(w(i), w(i+1)) \in J_M^h(K)$) for any $i = 1, \dots, m-1$. Furthermore define

$$U_M(w, K) = \bigcup_{v \in \Gamma_M(w,K)} K_v$$

for any $w \in T$.

Remark Define a weight function $h_* : T \to (0, 1]$ as $h_*(w) = 2^{-|w|}$ for any $w \in T$. Then $(T)_m = \Lambda_{2^{-m}}^{h_*}$ and $\Gamma_M(w, K) = \Lambda_{2^{-|w|}, M}^{h_*}(w)$.

Remark Note that the horizontal vertices E_m^h defined in Definition 2.2.11 is equal to $J_{1,m}^h$. On the contrary the collection $J_1^v(K)$ of the vertices in the vertical direction in $J_1(K)$ is strictly larger than that of vertical edges of the resolution $(T, \mathcal{B})$ in general.

If no confusion may occur, we are going to omit K in $\Gamma_M(w, K)$, $U_M(w, K)$, $J_{M,n}^h(K)$, $J_M^v(K)$, $J_M^h(K)$ and $J_M(K)$ and write $\Gamma_M(w)$, $U_M(w)$, $J_{M,n}^h$, J_M^v, J_M^h and J_M respectively. Moreover, in such a case, we simply say a (horizontal) M-jpath instead of a (horizontal) M-jpath of K.

Definition 4.1.2

(1) For $w \in T$ and $M \in \mathbb{N}$, define

$$C_w^M = \{(w(1), \dots, w(m)) | \pi(w(i)) \in \Gamma_M(w) \text{ for any } i = 1, \dots, m,$$

$$\text{there exist } w(0) \in S(w) \text{ and } w(m+1) \in (T)_{|w|+1} \backslash S(\Gamma_M(w))$$

$$\text{such that } (w(0), w(1), \dots, w(m), w(m+1)) \text{ is a horizontal } M\text{-jpath.}\}$$

(2) A function $\varphi : T \to (0, \infty)$ is called M-balanced if and only if

$$\sum_{i=1}^m \varphi(w(i)) \geq \varphi(\pi(w(m)))$$

for any $w \in T$ and $(w(1), \dots, w(m)) \in C_w^M$.

Remark If $J_M^h = \emptyset$, then $C_w^M = \emptyset$ for any $w \in T$ as well. Therefore, in this case, every $\varphi : T \to (0, \infty)$ is M-balanced. This happens if and only if the original set is (homeomorphic to) the Cantor set.

Theorem 4.1.3 *Define* $h_* : T \to (0, 1]$ *by* $h_*(w) = 2^{-|w|}$. *Let* $g \in \mathcal{G}_e(T)$. *Assume that* $g \underset{\text{GE}}{\sim} h_*$. *If there exists* $\varphi : T \to (0, \infty)$ *such that* φ *is* M-*balanced and* $\varphi \underset{\text{BL}}{\sim} g$, *i.e. there exist* $c_1, c_2 > 0$ *such that*

$$c_1 g(w) \le \varphi(w) \le c_2 g(w)$$

for any $w \in T$, *then there exists a metric* $\rho \in \mathcal{D}(X, \mathcal{O})$ *which is* M-*adapted to* g.

The rest of this section is devoted to a proof of the above theorem. Throughout this section, g and φ are assumed to satisfy the conditions required in Theorem 4.1.3.

To begin with, we summarize useful facts on g following immediately from the assumptions on g.

Proposition 4.1.4 *Let* g *be a weight function. Assume that* $g \underset{\text{GE}}{\sim} h_*$ *and that* g *is exponential.*

(1) *There exist* $\kappa \in (0, 1)$ *and* $N_0 \in \mathbb{N}$ *such that if* $|w| = |v|$ *and* $K_w \cap K_v \ne \emptyset$, *then*

$$g(w) \ge \kappa g(v) \tag{4.1.1}$$

and if $w, v \in \Lambda_s^g$ *and* $K_w \cap K_v \ne \emptyset$, *then*

$$|w| \le |v| + N_0. \tag{4.1.2}$$

(2) *There exist* $\eta \in (0, 1)$ *and* n_0 *such that*

$$\eta g(\pi(w)) \le g(w) \quad and \quad g(v) \le \eta g(w) \tag{4.1.3}$$

if $w \in T$, $v \in T_w$ *and* $|v| \ge |w| + n_0$. *Moreover,*

$$\eta s < g(w) \le s \tag{4.1.4}$$

for any $s \in (0, 1]$ *and* $w \in \Lambda_s^g$. *In other words, if* $g(w) \le \eta s$, *then* $\pi_g^*(w) \in \Lambda_t^g$ *for some* $t < s$.

Remark Recall that if $g(\pi^{n+1}(w)) > g(\pi^n(w)) = \cdots = g(\pi(w)) = g(w)$, then $\pi_g^*(w) = \pi^n(w)$ by Definition 3.5.3. Hence if $g(\pi(w)) > g(w)$, then $\pi_g^*(w) = w$.

The constants κ, N_0, n_0 and η are fixed in the rest of this section.

Lemma 4.1.5 *Assume that* $\varphi : T \to (0, \infty)$ *is* M-*balanced. Let* $m \ge 2$ *and let* $\mathbf{p} = (w(1), \ldots, w(m))$ *be an* M-*jpath satisfying* $|w(1)| = |w(m)| = |w(i)| - 1$ *for any* $i = \{2, \ldots, m - 1\}$. *Then* $w(m) \in \Gamma_M(w(1))$ *or there exists a horizontal*

M-jpath $\mathbf{p}' = (v(1), \ldots, v(n))$ such that $w(1) = v(1)$, $w(m) = v(n)$ and

$$\sum_{i=2}^{m-1} \varphi(w(i)) \geq \sum_{j=2}^{n-1} \varphi(v(j)).$$

Proof Assume that $w(m) \notin \Gamma_M(w(1))$. This implies that $m \geq 3$. We use an induction on m. Let $m \geq 3$. Since $w(1) \in \pi(\Gamma_M(w(2)))$, we have $\pi(w(2)) \in \Gamma_M(w(1))$.

Case 1: Suppose $\pi(w(2)), \ldots, \pi(w(m-1)) \in \Gamma_M(w(1))$.
In this case, since $(w(m-1), w(m)) \in J_M^v$, there exists $u \in \Gamma_M(w(m-1))$ such that $\pi(u) = w(m)$. By the fact that $\pi(u) = w(m) \notin \Gamma_M(w(1))$, we confirm $(w(2), \ldots, w(m-1)) \in \mathcal{C}_{w(1)}^M$. Since φ is M-balanced, we see

$$\varphi(w(2)) + \cdots + \varphi(w(m-1)) \geq \varphi(\pi(w(m-1))).$$

Hence $(w(1), \pi(w(m-1)), w(m))$ is the desired horizontal M-jpath. Note that if $m = 3$, then Case 1 applies.

Case 2: Suppose that $\pi(w(2)), \ldots, \pi(w(i)) \in \Gamma_M(w(1))$ and $\pi(w(i+1)) \notin \Gamma_M(w(1))$ for some $i \in \{2, \ldots, m-2\}$.
In this case, set $\widetilde{p} = (w(1), \pi(w(i)), w(i+1), \ldots, w(m))$. Since $(w(1), w(2)) \in J_M^v$, there exists $v \in T$ such that $\pi(v) = w(1)$ and $v \in \Gamma_M(w(2))$. Therefore, $(v, w(2), \ldots, w(i), w(i+1))$ is a horizontal M-jpath and $(w(2), \ldots, w(i)) \in \mathcal{C}_{w(1)}^M$. Since φ is M-balanced, it follows that

$$\varphi(w(2)) + \cdots + \varphi(w(i)) \geq \varphi(\pi(w(i))).$$

Therefore

$$\sum_{k=2}^{m-1} \varphi(w(k)) \geq \varphi(\pi(w(i))) + \sum_{j=i+1}^{m-1} \varphi(w(j)).$$

Applying induction hypothesis to $(\pi(w(i)), w(i+1), \ldots, w(m-1), w(m))$, we obtain the desired result in this case. □

Repeated use of the above lemma yields the following fact.

Lemma 4.1.6 Assume that φ is M-balanced. Let $\mathbf{p} = (w(1), \ldots, w(m))$ is an M-jpath satisfying $|w(1)| = |w(m)|$. Then $w(m) \in \Gamma_M(w(1))$ or there exists an M-jpath $\mathbf{p}' = (v(1), \ldots, v(k))$ such that $v(1) = w(1)$, $v(k) = w(m)$, $|v(i)| \leq |v(1)|$

for any $i = 1, \ldots, k$, $k \leq m$ and

$$\sum_{i=2}^{m-1} \varphi(w(i)) \geq \sum_{j=2}^{k-1} \varphi(v(j)).$$

Lemma 4.1.7 *Assume that φ is M-balanced. Let $\mathbf{p} = (w(1), \ldots, w(m))$ be an M-jpath. Suppose that $w, v \in T$, $|w| = |v|$, $w \notin \Gamma_M(v)$, $w(1) \in T_w$ and $w(m) \in T_v$.*

(1) *There exists an M-jpath $(v(1), \ldots, v(k))$ such that $v(1) = w, w(k) = v$, $|v(j)| \leq |w|$ for any $j = 1, \ldots, k$ and*

$$\sum_{i=2}^{m-1} \varphi(w(i)) \geq \sum_{j=2}^{k-1} \varphi(v(j)).$$

(2) *Assume that there exists $\kappa_0 \in (0, 1)$ such that $\varphi(w) \geq \kappa_0 \varphi(v)$ for any $(w, v) \in J_M$. Then*

$$\sum_{i=2}^{m-1} \varphi(w(i)) \geq \kappa_0 \max\{\varphi(w), \varphi(v)\}.$$

Proof (1) Since $w(1) \in T_w$ and $w(m) \in T_v$, there exist $n_1, n_2 \geq 0$ such that $\pi^{n_1}(w(1)) = w$ and $\pi^{n_2}(w(m)) = v$. We use an induction on $n_1 + n_2$. If $n_1 + n_2 = 0$, then Lemma 4.1.6 suffices. Assume that $n_1 + n_2 \geq 1$. Suppose that there exists $(w(k), w(k + 1), \ldots, w(k + l))$ such that $|w(k)| = |w(k + l)|$ and $|w(k + i)| = |w(k)| + 1$ for any $i = 1, \ldots, l - 1$. We call such a sequence $(w(k), \ldots, w(k + l))$ as a plateau. If $w(k + l) \in \Gamma_M(w(k))$, then we replace this part by $(w(k), w(k + l))$. Otherwise using Lemma 4.1.5, we can replace it by a horizontal M-jpath $(w(k), w'(1), \ldots, w'(k'), w(k + l))$ satisfying

$$\sum_{i=1}^{l-1} \varphi(w(k + i)) \geq \sum_{j=1}^{k'} \varphi(w'(j)).$$

Making iterated use of this procedure, we may assume that $(w(1), \ldots, w(m))$ contains no plateau without loss of generality. Suppose $n_1 \geq n_2$. Then $n_1 \geq 1$. If $|w(1)| + 1 = |w(2)|$, then no plateau assumption yields $|w(1)| < |w(2)| \leq \ldots \leq |w(m)|$. Therefore, we have $n_1 < n_2$ and this contradicts the fact that $n_1 \geq n_2$. Hence we see $|w(1)| \geq |w(2)|$. Then $(\pi(w(1)), w(2)) \in J_M$. Hence $(\pi(w(1)), w(2), \ldots, w(m))$ is an M-jpath to which the induction hypothesis applies. If $n_1 \leq n_2$, then same argument works by replacing $w(1)$ with $w(m)$. Thus we have shown (1).

(2) By (1), there exists an M-jpath $(v(1), \ldots, v(k))$ such that $v(1) = w$, $v(k) = v$, $|v(i)| \leq |w|$ for any $i = 1, \ldots, k$ and

$$\sum_{i=2}^{m-1} \varphi(w(i)) \geq \sum_{j=2}^{k-1} \varphi(v(j)).$$

If $k = 3$, we have $(w, v(2)), (v(2), v) \in J_M$. Hence

$$\sum_{i=2}^{m-1} \varphi(w(i)) \geq \varphi(v(2)) \geq \kappa_0 \max\{\varphi(w), \varphi(v)\}.$$

If $k > 3$, then $(w, v(1)), (v(k-1), v) \in J_M$ and hence

$$\sum_{i=2}^{m-1} \varphi(w(i)) \geq \varphi(v(2)) + \varphi(v(k-1)) \geq \kappa_0(\varphi(w) + \varphi(v)) \geq \kappa_0 \max\{\varphi(w), \varphi(v)\}.$$

$\square$

Definition 4.1.8 Let $\varphi : T \to (0, \infty)$.

(1) For an M-jpath $\mathbf{p} = (w(1), \ldots, w(m))$, we define

$$\ell_M^\varphi(\mathbf{p}) = \sum_{i=1}^{m} \varphi(w(i)).$$

(2) For a chain $\mathbf{p} = (w(1), \ldots, w(m))$ of K, define $L_\varphi(\mathbf{p})$ by

$$L_\varphi(\mathbf{p}) = \sum_{i=1}^{m} \varphi(w(i)).$$

Since there is no jump in a 1-jpath, we say 1-path instead of 1-jpath in the followings.

Lemma 4.1.9 Let $g \in \mathcal{G}_e(T)$. For any chain $\mathbf{p} = (w(1), \ldots, w(m))$ of K, there exists a 1-path $\widehat{\mathbf{p}} = (v(1), \ldots, v(k))$ of K such that $w(1) = v(1)$, $w(m) = v(k)$ and

$$L_g(\mathbf{p}) \geq c\ell_1^g(\widehat{\mathbf{p}}),$$

where $c > 0$ is independent of $\mathbf{p}$ and $\widehat{\mathbf{p}}$.

Proof By (4.1.3), there exist $c \geq 1$ and $\lambda \in (0, 1)$ such that $g(w) \leq c\lambda^k g(\pi^k(w))$ for any $w \in T$ and $k \geq 0$.

Now we start to construct a 1-path $\widehat{\mathbf{p}}$ by inserting a sequence between $w(i)$ and $w(i+1)$ for each i with $|w(i)| \neq |w(i+1)|$. If $|w(i)| > |w(i+1)|$, then there exists

$v \in T_{w(i+1)}$ such that $|v| = |w(i)|$ and $K_{w(i)} \cap K_v \neq \emptyset$. Let $k_i = |w(i)| - |w(i+1)|$. Then $\pi^{k_i}(v) = w(i+1)$ and $(w(i), v, \pi(v), \dots, \pi^{k_i}(v))$ is a 1-path. Moreover,

$$\sum_{j=0}^{k_i-1} g(\pi^j(v)) \leq (c\lambda^{k_i} + c\lambda^{k_i-1} + \dots + c\lambda)g(w(i+1)) \leq \frac{c\lambda}{1-\lambda}g(w(i+1)).$$

$$(4.1.5)$$

Next suppose that $|w(i)| < |w(i+1)|$. Then using a similar discussion as above, we find a 1-path $(\pi^{k_i}(v), \pi^{k_i-1}(v), \dots, v, w(i+1))$ satisfying $\pi^{k_i}(v) = w(i)$ and a counterpart of (4.1.5). Inserting sequences in this manner, we obtain the desired $\widehat{\mathbf{p}}$. By (4.1.5), it follows that

$$\left(1 + \frac{2c\lambda}{1-\lambda}\right)L_g(\mathbf{p}) \geq \ell_1^g(\widehat{\mathbf{p}}).$$

$\square$

Proof of Theorem 4.1.3 Fix $M \in \mathbb{N}$. Write $\delta(x, y) = \delta_M^g(x, y)$ for any $x, y \in X$. For $A \subseteq X$, we set

$$\Lambda_s^g(A) = \{w | w \in \Lambda_s^g, K_w \cap A \neq \emptyset\},$$

$$(T)_n(A) = \{w | w \in (T)_n, K_w \cap A \neq \emptyset\}.$$

In particular, we write $\Lambda_s^g(x, y) = \Lambda_s^g(\{x, y\})$ and $(T)_n(x, y) = (T)_n(\{x, y\})$.

Since $\varphi \underset{\mathrm{BL}}{\sim} g$, using Proposition 4.1.4, we see that there exists $\kappa_0 \in (0, 1)$ such that $\varphi(w) \geq \kappa_0 \varphi(v)$ for any $(w, v) \in J_M$.

Claim 1 There exists $N_1 \in \mathbb{N}$ such that

$$||w| - |v|| \leq N_1$$

for any $x, y \in X$ and $w, v \in \Lambda_{\kappa M \delta(x,y)}^g(x, y)$.

Proof of Claim 1 By (4.1.3), there exists $N' \in \mathbb{N}$ such that if $w \in \Lambda_{\kappa M_s}^g$, $w' \in \Lambda_s^g$ and $w \in T_{w'}$, then $|w| \leq |w'| + N'$. Let $w, v \in \Lambda_{\kappa M \delta(x,y)}^g(x, y)$. If $w, v \in \Lambda_{\kappa M \delta(x,y)}^g(x)$ or $w, v \in \Lambda_{\kappa M \delta(x,y)}^g(y)$, then we have $K_w \cap K_v \neq \emptyset$ and hence $||w| - |v|| \leq N_0$ by (4.1.2). Otherwise, we may assume that $x \in K_w$ and $y \in K_v$ without loss of generality. Choose $w', v' \in \Lambda_{\delta(x,y)}^g(x, y)$ so that $w \in T_{w'}$ and $v \in T_{v'}$. Since $y \in U_M^g(x, \delta(x, y))$ by Proposition 2.3.9, there exists a chain $(w(1), \dots, w(M+3))$ such that $w(1) = w'$, $w(M+3) = v'$ and $w(j) \in \Lambda_{\delta(x,y)}^g$ for any $j = 1, \dots, M+3$. By (4.1.2), it follows that

$$|w| - N' \leq |w'| \leq |v'| + N_0(M+2) \leq |v| + N_0(M+2).$$

Hence letting $N_1 = N_0(M+2) + N'$, we obtain the desired claim. $\square$

Claim 2 For any $N \in \mathbb{N}$, if $w \in \Lambda_s^g$, $v \in \Lambda_{\eta^{N+1}s}^g$ and $v \in T_w$, then $|v| \geq |w| + N$.

Proof of Claim 2 By (4.1.3) and (4.1.4), it follows that

$$\eta^{|v|-|w|+1}s \leq \eta^{|v|-|w|}g(w) \leq g(v) \leq \eta^{N+1}s.$$

Hence $|v| - |w| \geq N$. $\square$

Claim 3 Set $N_2 = N_1 + 2$ and $r_* = \eta^{N_2}\kappa^M$. For any $x, y \in X$, there exists $l_* = l_*(x, y) \in \mathbb{N}$ such that $|w| \geq l_*$ for any $w \in \Lambda_{r_*\delta(x,y)}^g(x, y)$ and if $v \in (T)_{l_*}(x, y)$, then $g(v) \leq \kappa^M\delta(x, y)$.

Proof of Claim 3 Set $s = \kappa^M\delta(x, y)$. Let

$$N_3 = \min_{u \in \Lambda_s^g(x,y)} |u|.$$

Set $l_* = l_*(x, y) = N_3 + N_1 + 1$. Note that $r_*\delta(x, y) = \eta^{N_2}s < s$. For any $w \in \Lambda_{r_*\delta(x,y)}^g(x, y)$, choose $w_* \in \Lambda_s^g(x, y)$ so that $w \in T_{w_*}$. By Claims 1 and 2,

$$|w| \geq |w_*| + N_1 + 1 \geq N_3 + N_1 + 1 = l_* > N_3 + N_1 \geq |u|$$

for any $u \in \Lambda_s^g(x, y)$. Let $v \in (T)_{l_*}(x, y)$. There exists $v' \in \Lambda_s^g(x, y)$ such that $v \in T_{v'}$. Since $|v'| < l_*$, it follows that $g(v) \leq g(v') \leq s$. $\square$

Let $\mathbf{p} = (w(1), \dots, w(m))$ be a chain of K. Assume that $x \in K_{w(1)}$ and $y \in K_{w(m)}$. If $g(w(1)) \geq r_*\delta(x, y)$ or $g(w(m)) \geq r_*\delta(x, y)$, then

$$L_g(\mathbf{p}) \geq r_*\delta(x, y). \tag{4.1.6}$$

Assume that $g(w(1)) < r_*\delta(x, y)$ and $g(w(m)) < r_*\delta(x, y)$. Set $l_* = l_*(x, y)$. Then by Claim 3, there exist $w, v \in (T)_{l_*}$ such that $w(1) \in T_w$, $w(m) \in T_v$, $\max\{g(w), g(v)\} \leq \kappa^M\delta(x, y)$. Note that $x \in K_w$ and $y \in K_v$. Suppose that $v \in \Gamma_M(w)$. Then there exists a horizontal M-chain $(u(1), \dots, u(M+1))$ such that $u(1) = w$ and $u(M+1) = v$. Let $s_* = \max\{g(u(i))|i = 1, \dots, M+1\}$. Then $(\pi^{m_1}(u(1)), \dots, \pi^{m_{M+1}}(u(M+1)))$ is a M-chain between x and y in $\Lambda_{s_*}^g$ for some $m_1, \dots, m_{M+1} \geq 0$. Therefore, $\delta(x, y) \leq s_*$. On the other hand, since $\max\{g(w), g(v)\} \leq \kappa^M\delta(x, y)$, by (4.1.1) we see that

$$g(u(i)) \leq \kappa^{-\max\{i-1, M-i+1\}}\max\{g(w), g(v)\} \leq \kappa\delta(x, y) < \delta(x, y)$$

for any $i = 1, \dots, M+1$. This implies that $s_* < \delta(x, y)$. Thus it follows that $v \notin \Gamma_M(w)$. By Lemma 4.1.9, there exists a 1-path $\mathbf{p}_1 = (v(1), \dots, v(k))$ such that

$v(1) = w(1)$, $v(k) = w(m)$ and

$$L_g(\mathbf{p}) \geq c_0 \ell_1^g(\mathbf{p}_1), \tag{4.1.7}$$

where c_0 is independent of $\mathbf{p}$. Note that a 1-path of K is an M-jpath of K for any $M \geq 1$ and $\ell_M^g(\mathbf{p}_1) = \ell_1^g(\mathbf{p}_1)$. Since $\varphi \underset{\mathrm{BL}}{\sim} g$, we have

$$\ell_M^g(\mathbf{p}_1) \geq c_1 \ell_M^\varphi(\mathbf{p}_1), \tag{4.1.8}$$

where $c_1 > 0$ is independent of $\mathbf{p}$. Applying Lemma 4.1.7, we obtain

$$\ell_M^\varphi(\mathbf{p}_1) \geq \kappa_0 \max\{\varphi(w), \varphi(v)\} \geq c_2 \kappa_0 \max\{g(w), g(v)\}, \tag{4.1.9}$$

where c_2 is independent of $\mathbf{p}$. By Claim 3, there exist $w' \in T_w$ and $v' \in T_v$ such that $w', v' \in \Lambda_{r_*\delta(x,y)}^g$. Hence by (4.1.4), we have $\eta r_* \delta(x, y) < g(w') \leq g(w)$ and $\eta r_* \delta(x, y) < g(v') \leq g(v)$. So by (4.1.9),

$$\ell_M^\varphi(\mathbf{p}_1) \geq c_3 \delta(x, y), \tag{4.1.10}$$

where c_3 is independent of $\mathbf{p}$. Finally combining (4.1.6)–(4.1.8) and (4.1.10), we conclude that there exists $c_4 > 0$ such that if $\mathbf{p} = (w(1), \dots, w(m))$ is a chain of K, $x \in K_{w(1)}$ and $y \in K_{w(m)}$, then

$$L_g(\mathbf{p}) \geq c_4 \delta(x, y).$$

This immediately implies

$$c_4 \delta_M^g(x, y) \leq D^g(x, y) \leq D_M^g(x, y) \leq (M + 1)\delta_M^g(x, y)$$

for any $x, y \in X$. Thus, the metric D^g is M-adapted to g. □

4.2 Construction of Ahlfors Regular Metric I

In this section, we discuss a condition for a weight function to induce an Ahlfors regular metric, whose definition is given in Definition 4.2.1.

As in the last section, $(X, \mathcal{O})$ is a compact metrizable topological space with no isolated point, $(T, \mathcal{A}, \phi)$ is a locally finite tree and $K : T \to \mathcal{C}(X, \mathcal{O})$ is a minimal partition. Furthermore, we assume that the partition $K : T \to \mathcal{C}(X, \mathcal{O})$ is strongly finite, i.e. $\sup_{w \in T} \#(S(w)) < +\infty$ throughout this section.

Definition 4.2.1 A metric $d \in \mathcal{D}(X, \mathcal{O})$ is called α-Ahlfors regular if there exist a Borel regular probability measure μ on X such that μ is α-Ahlfors regular with respect to d. Furthermore d is called Ahlfors regular if d is α-Ahlfors regular for some $\alpha > 0$.

Theorem 4.2.2 *Let $d \in \mathcal{D}_{A,e}(X, \mathcal{O})$ and assume that $d \underset{GE}{\sim} h_*$, d is thick and uniformly finite. Let $\alpha > 0$. There exist a metric $\rho \in \mathcal{D}(X, \mathcal{O})$ and a measure $\mu \in \mathcal{M}_P(X, \mathcal{O})$ such that $\rho \underset{QS}{\sim} d$ and μ is α-Ahlfors regular with respect to ρ if and only if there exists $g \in \mathcal{G}_e(X)$ such that*

- *$g \underset{GE}{\sim} g_d$,*
- *there exist $c > 0$ and $M \geq 1$ such that*

$$cD_M^g(x, y) \leq D^g(x, y) \tag{4.2.1}$$

for any $x, y \in X$,
- *there exists $c > 0$ such that*

$$c^{-1}g(w)^\alpha \leq \sum_{v \in S^n(w)} g(v)^\alpha \leq cg(w)^\alpha \tag{4.2.2}$$

for any $w \in T$ and $n \geq 0$, where $S^n(w) = (T)_{|w|+n} \cap T_w$ by definition.

In particular, if $\rho \in \mathcal{D}(X, O)$, $\mu \in \mathcal{M}_P(X, \mathcal{O})$, $\rho \underset{QS}{\sim} d$ and μ is α-Ahlfors regular with respect to ρ, then (4.2.2) holds with $g = g_\rho$.

Remark The condition (4.2.1) is equivalent to the existence of a metric ρ' which is M-adapted to g for some $M \geq 1$.

Proof By Proposition 3.1.7, d is tight. Hence Theorem 3.5.7 shows that h_* is tight, thick and uniformly finite.

Suppose that there exist a metric ρ and a measure μ such that $\rho \underset{QS}{\sim} d$ and μ is α-Ahlfors regular with respect to ρ. By Theorem 3.6.6, setting $g = g_\rho$, we see that $g \underset{GE}{\sim} g_d$, g is exponential and ρ is adapted. Hence by Theorem 3.5.7, g is thick and uniformly finite. Using Proposition 2.4.8, we verify (4.2.1). Since μ has the volume doubling property with respect to ρ, Lemma 3.3.11 implies that

$$\sum_{v \in S^n(w)} c\mu(K_v) \leq \sum_{v \in S^n(w)} \mu(O_v) \leq \mu(K_w) \leq \sum_{v \in S^n(w)} \mu(K_v).$$

Furthermore, by Theorem 3.1.21, it follows that $g^\alpha \underset{BL}{\sim} g_\mu$. This yields (4.2.2).

Conversely, assume that $g \in \mathcal{G}_e(X)$, $g \underset{GE}{\sim} g_d$, (4.2.1) and (4.2.2). Since $g, g_d \in \mathcal{G}_e(T)$, $g \underset{GE}{\sim} g_d$ and g_d is tight and thick, Theorem 3.5.7 shows that g is thick and tight. Define $\rho(x, y) = D^g(x, y)/\sup_{a,b \in X} D^g(a, b)$. By (4.2.1), it follows that ρ is adapted to g. Moreover, Theorem 3.1.14 implies that $g_\rho \underset{BL}{\sim} g$. Therefore $g_\rho \underset{GE}{\sim} g_d$

and hence by Theorem 3.6.6, we see that $\rho \underset{QS}{\sim} d$. Choose $x_w \in O_w$ for each $w \in T$.
Define

$$\mu_n = \frac{1}{\sum_{w\in(T)_n} g(w)^\alpha} \sum_{w\in(T)_n} g(w)^\alpha \delta_{x_w},$$

where δ_x is Dirac's point mass at x. Note that (4.2.2) implies that

$$c^{-1} = c^{-1}g(\phi) \leq \sum_{w\in(T)_n} g(w)^\alpha \leq cg(\phi) = c. \tag{4.2.3}$$

Since $(X, \mathcal{O})$ is compact, there exist a sub-sequence $\{\mu_{n_i}\}_{i\geq 1}$ and a Borel regular
probability measure μ on X such that $\{\mu_{n_i}\}_{i\geq 1}$ converges weakly to μ as $i \to \infty$.
For $w \in T$, let $A_{w,\epsilon}$ be the ϵ-neighborhood of K_w with respect to d, i.e. $A_{w,\epsilon} = \{y | y \in X, \inf_{x\in K_w} d(x, y) < \epsilon\}$ and let $f_{w,\epsilon} : X \to [0, 1]$ be a continuous function
satisfying $f_{w,\epsilon}|_{K_w} = 1$ and $f_{w,\epsilon}|_{X\setminus A_{w,\epsilon}} = 0$. Set

$$U_w = \mathrm{int}\left(\bigcup_{v\in\Gamma_1(w)} K_v \right).$$

Then U_w is an open neighborhood of K_w. Therefore for sufficiently small $\epsilon > 0$,
$A_{w,\epsilon} \subseteq U_w$. By this fact, we have

$$\sum_{v\in S^n(w)} g(v)^\alpha \leq \int_X f_{w,\epsilon} d\mu_{|w|+n} \leq \sum_{u\in\Gamma_1(w)} \sum_{v\in S^n(u)} g(v)^\alpha.$$

By (4.2.2),

$$c^{-1}g(w)^\alpha \leq \int_X f_{w,\epsilon} d\mu_{|w|+n} \leq c \sum_{u\in\Gamma_1(w)} g(u)^\alpha.$$

Since h_* is uniformly finite, there exists $c_1 > 0$, which is independent of w, such
that $\#(\Gamma_1(w)) \leq c_1$. Moreover since $g \underset{GE}{\sim} d \underset{GE}{\sim} h_*$, using Proposition 4.1.4, we have

$$c^{-1}g(w)^\alpha \leq \int_X f_{w,\epsilon} d\mu_{|w|+n} \leq cc_1\kappa^{-\alpha}g(w)^\alpha.$$

Choosing the proper subsequence of $|w| + n$ and taking the limit, we obtain

$$c^{-1}g(w)^\alpha \leq \int_X f_{w,\epsilon} d\mu \leq cc_1\kappa^{-\alpha}g(w)^\alpha.$$

Letting $\epsilon \downarrow 0$, we see

$$c^{-1}g(w)^{\alpha} \leq \mu(K_w) \leq cc_1\kappa^{-\alpha}g(w)^{\alpha}.$$

This implies that $\mu \in \mathcal{M}_P(X, \mathcal{O})$ and $g^{\alpha} \underset{\text{BL}}{\sim} g_{\mu}$. Moreover by Theorem 3.5.7, g_{ρ} is uniformly finite. Hence by Theorem 3.1.21, μ is α-Ahlfors regular with respect to d.   □

4.3 Basic Framework

From this section, we start proceeding towards the characterization of Ahlfors regular conformal dimension. To begin with, we fix our framework in this section and are going to keep it until the end.

As in the previous sections, $(T, \mathcal{A}, \phi)$ is a locally finite tree with the root ϕ, $(X, \mathcal{O})$ is a compact metrizable topological space with no isolated point, $K : T \to \mathcal{C}(X, \mathcal{O})$ is a minimal partition. We also assume that $(T, \mathcal{A}, \phi)$ is strongly finite, i.e. $\sup_{w \in T} \#(S(w)) < +\infty$.

Our standing assumptions in the following sections are as follows:

Basic Framework Let $d \in \mathcal{D}(X, \mathcal{O})$. For $r \in (0, 1)$, define $h_r : T \to (0, 1]$ by

$$h_r(w) = r^{|w|}$$

for any $w \in T$. We assume the following conditions (BF1) and (BF2) are satisfied:

(BF1) d is M_*-adapted for some $M_* \geq 1$, exponential, thick, and uniformly finite.
(BF2)   There exists $r \in (0, 1)$ such that $h_r \underset{\text{BL}}{\sim} d$.

Remark Note that h_r is an exponential weight function. Hence the condition that d being exponential in (BF1) can be omitted if we assume (BF2).

Remark Since d is exponential by (BF1), Lemma 3.6.4 show that (X, d) is uniformly perfect.

Remark Under (BF1) and (BF2), d is M_*-adapted to h_r and h_r is tight, thick and uniformly finite.

Our goal of the rest of this monograph is to obtain characterizations of the Ahlfors regular conformal dimension of (X, d) satisfying the above conditions (BF1) and (BF2).

The condition (BF2) may be too restrictive. Replacing the original tree T by its subtree $\widetilde{T}^{g_d, r}$ associated with d defined in Definition 2.5.10, however, we can realize (BF2) providing (BF1) is satisfied.

Now assuming (BF1), since d is exponential, there exists $\eta \in (0, 1)$ such that $g(w) \geq \eta g(\pi(w))$ for any $w \in T$. We use this constant η through this section.

To simplify the notation, we write $\widetilde{T}^{d,r}$ in place of $\widetilde{T}^{g_d,r}$ hereafter.

Proposition 4.3.1 *Assume that d is M_*-adapted, exponential, thick and uniformly finite. For any $r \in (0, \eta]$, if we replace T and $K : T \to \mathcal{C}(X, \mathcal{O})$ by $\widetilde{T}^{d,r}$ and $K_{\widetilde{T}^{d,r}} : \widetilde{T}^{d,r} \to \mathcal{C}(X, \mathcal{O})$ respectively, then* (BF1) *and* (BF2) *are satisfied.*

Proof Since we should handle two different structures (T, K) and $(\widetilde{T}^{d,r}, K_{\widetilde{T}^{d,r}})$ here, we denote Λ_s^g and $U_M^g(x, s)$ by $\Lambda_s^g(T, K)$ and $U_M^g(x, s; T, K)$ respectively to emphasize the dependency on a tree structure T and a partition K. Moreover, for ease the notation, we set

$$\widetilde{T} = \widetilde{T}^{g_d,r}, \ \widetilde{K} = K_{\widetilde{T}^{g_d,r}}, \ \pi = \pi^{(T,\mathcal{A},\phi)}, \ \widetilde{\pi} = \pi^{(\widetilde{T},\widetilde{\mathcal{A}},\phi)},$$

where $\widetilde{\mathcal{A}}$ is the natural tree structure of $\widetilde{T}$, and

$$\Lambda_s^g = \Lambda_s^g(T, K), U_M^g(x, s) = U_M^g(x, s; T, K),$$
$$\widetilde{\Lambda}_s^g = \Lambda_s^g(\widetilde{T}, \widetilde{K}), \widetilde{U}_M^g(x, s) = U_M^g(x, s; \widetilde{T}, \widetilde{K}).$$

Since $r \leq \eta$, if $d(\pi(w)) > r^n \geq d(w)$, then $d(w) > r^{n+1}$. Hence $\Lambda_{r^n}^d \cap \Lambda_{r^{n+1}}^d = \emptyset$ and, for any $w \in \widetilde{T} \backslash \{\phi\}$, $\widetilde{\pi}(w) = \pi^k(w)$ for some $k \geq 1$. This implies, for any $n \geq 0$,

$$\widetilde{\Lambda}_{r^n}^{\widetilde{h}_r} = (\widetilde{T})_n = \Lambda_{r^n}^d = \widetilde{\Lambda}_{r^n}^d,$$

where $\widetilde{h}_r : \widetilde{T} \to (0, 1]$ is defined by $\widetilde{h}_r(w) = r^n$ for any $w \in (\widetilde{T})_n$. Consequently,

$$U_{M_*}^d(x, r^n) = \widetilde{U}_{M_*}^{\widetilde{h}_r}(x, r^n).$$

Since d is M_*-adapted, there exist $c_1, c_2 > 0$ such that

$$B_d(x, c_1 s) \subseteq U_{M_*}^d(x, s) \subseteq B_d(x, c_2 s) \tag{4.3.1}$$

for any $x \in X$ and $s \in (0, 1]$. Let $r^m > s \geq r^{m+1}$. Then

$$\widetilde{U}_{M_*}^{\widetilde{h}_r}(x, s) \subseteq \widetilde{U}_{M_*}^{\widetilde{h}_r}(x, r^m) = U_{M_*}^d(x, r^m) \subseteq B_d(x, c_2 r^m) \subseteq B_d(x, c_2 r^{-1} s).$$

In the same manner, we also obtain

$$B_d(x, c_1 r s) \subseteq \widetilde{U}_{M_*}^{\widetilde{h}_r}(x, s).$$

Thus d is M_*-adapted with respect to $(\widetilde{T}, \widetilde{K})$.

Exponential Property of d Let $w \in (\widetilde{T})_n$. Since $\widetilde{\pi}(w) \in (\widetilde{T})_{n-1}$, we see that $r^{n-1} \geq d(\widetilde{\pi}(w)) > r^n \geq d(w) > r^{n+1}$. This shows that $d(w) \geq r^2 d(\widetilde{\pi}(w))$ and so d is super-exponential with respect to $\widetilde{T}$. Sub-exponential property of d follows from the fact that $\widetilde{\pi}(w) = \pi^k(w)$ for some $k \geq 1$.

Thickness of d Let $w \in \widetilde{\Lambda}_s^d$. Then $d(\widetilde{\pi}(w)) > s \geq d(w)$. If $w \in (\widetilde{T})_n$, then $d(\pi(w)) > r^n \geq d(w)$. Since $d(w) \geq r^2 d(\widetilde{\pi}(w))$, we see that $r^n \geq r^2 s$. On the other hand, since d is thick with respect to (T, K), there exist $M \geq 1$ and $\alpha > 0$ such that if $w \in \Lambda_{r^n}^d$, then

$$K_w \supseteq U_M^d(x, \alpha r^n)$$

for some $x \in K_w$. Choose m so that $r^m < \alpha$. Then

$$K_w \supseteq U_M^d(x, \alpha r^n) \supseteq U_M^d(x, r^{m+n}) = \widetilde{U}_M^d(x, r^{m+n}) \supseteq \widetilde{U}_M^d(x, r^{m+2}s).$$

Thus d is thick with respect to $(\widetilde{T}, \widetilde{K})$.

Uniform Finiteness of d This is immediate from the fact that $\widetilde{\Lambda}_{r^n}^d = \Lambda_{r^n}^d$.

$d \underset{BL}{\sim} \widetilde{h}_r$ Since d is exponential, there exists $c > 0$ such that

$$cd(w) \geq s \geq d(w)$$

if $w \in \Lambda_s^d$. This implies that $\widetilde{h}_r \underset{BL}{\sim} d$ as a weight function of $\widetilde{T}$.

Thus (BF1) and (BF2) with respect to $(\widetilde{T}^{d,r}, K_{\widetilde{T}^{d,r}})$ are satisfied.   □

Due to this proposition, if d is M_*-adapted, exponential, thick and uniformly finite, then replacing the original tree $(T, \mathcal{A}, \phi)$ and the partition K by the tree $(\widetilde{T}^{d,r}, \widetilde{\mathcal{A}}, \phi)$ and the partition $K_{\widetilde{T}^{d,r}}$ respectively, we always assume that $(T, \mathcal{A}, \phi)$ and K satisfy (BF1) and (BF2) hereafter.

Note that even after the modification the condition $\sup_{w \in T} \#(S(w)) < +\infty$ still holds as d is exponential. Moreover, since d is uniformly finite, the following supremums are finite.

Definition 4.3.2 Define

$$L_* = \sup_{w \in T} \#(\Gamma_1(w))$$

and

$$N_* = \sup_{w \in T} \#(S(w)).$$

Notation We write $U_M(x, s) = U_M^{h_r}(x, s)$ for any $M \geq 0$, $x \in X$ and $s \in (0, 1]$.

By the above definition, it is straightforward to deduce the next lemma.

Lemma 4.3.3

(1) *For any $w \in T$ and $k \geq 1$,*

$$\#(S^k(w)) \leq (N_*)^k. \qquad (4.3.2)$$

(2) *For any $w \in T$ and $M \geq 1$,*

$$\#(\Gamma_M(w)) \leq (L_*)^M. \qquad (4.3.3)$$

We present two useful propositions. The first one is an observation on the geometry of $\Gamma_M(w)$'s which holds without (BF1) and (BF2).

Proposition 4.3.4 *Let $M_1, M_2 \in \mathbb{N}$. Suppose that*

$$\pi(\Gamma_{M_1+M_2}(v)) \subseteq \Gamma_{M_2}(\pi(v)) \qquad (4.3.4)$$

for any $v \in T$. Then

$$\pi(\Gamma_{M_1+M_2}(v)) \subseteq \Gamma_{M_1+M_2}(w)$$

if $\pi(v) \in \Gamma_{M_1}(w)$.

Proof Assume that $\pi(v) \in \Gamma_{M_1}(w)$, i.e. there exists a horizontal 1-jpath $(w(1), \ldots, w(M_1 + 1))$ such that $w(1) = w$ and $w(M_1 + 1) = \pi(v)$. By (4.3.4), we see $\pi(u) \in \Gamma_{M_2}(\pi(v))$ for any $u \in \Gamma_{M_1+M_2}(v)$. Hence there exists a horizontal 1-path $(v(1), \ldots, v(M_2 + 1))$ such that $v(1) = \pi(v)$ and $v(M_2 + 1) = \pi(u)$. As $(w(1), \ldots, w(M_1), v(1), \ldots, v(M_2 + 1))$ is a horizontal 1-path, we see that $\pi(u) \in \Gamma_{M_1+M_2}(w)$. $\qquad \square$

The second observation requires (BF1) and (BF2).

Proposition 4.3.5 *Assume (BF1) and (BF2). For any $M \geq 1$, there exists $m_0 \in \mathbb{N}$ such that, for any $m \geq m_0$ and $w \in T$, $\Gamma_M(v) \subseteq S^m(w)$ for some $v \in S^m(w)$ and $\#(S^m(w)) \geq 2$.*

To prove this proposition, we need the following fact.

Lemma 4.3.6 *Suppose that a partition K is minimal. Let $A \subseteq T$ and let $v \in T$. If $|v| \geq |w|$ for any $w \in A$ and $K_v \subseteq \cup_{w \in A} K_w$, then $v \in \cup_{w \in A} T_w$.*

Proof Set $A' = \cup_{w \in A} S^{|v|-|w|}(w)$. Then $A' \subseteq (T)_{|v|}$ and

$$K_v \subseteq \bigcup_{w \in A} K_w = \bigcup_{u \in A'} K_u.$$

Since $O_v \neq \emptyset$, we see that $v \in A' \subseteq \cup_{w \in A} T_w$. $\qquad \square$

Proof of Proposition 4.3.5 Since d is thick, so does h_r. By Proposition 3.2.1, there exists $\alpha \in (0, 1)$ such that for any $w \in T$,

$$U_M(x, \alpha r^{|w|}) \subseteq K_w$$

for some $x \in K_w$. Set $m_1 = \min\{m \mid r^m < \alpha\}$ and let $m \geq m_0$. Then $U_M(x, r^{|w|+m}) \subseteq K_w$. Therefore, if $v \in S^m(w)$ and $x \in K_v$, then $U_M(v, K) \subseteq U_M(x, r^{|w|+m}) \subseteq K_w$. Since the partition is minimal, Lemma 4.3.6 implies that $\Gamma_M(v) \subseteq S^m(w)$. Now, since d is exponential, there exist $\eta \in (0, 1)$ and $n \geq 1$ such that $d(v) \leq \eta d(w)$ if $v \in T_w$ and $|v| \geq |w| + n$. Since $d(v) = \mathrm{diam}(K_v, d) < d(w) = \mathrm{diam}(K_w, d)$, we see that $K_v \subsetneq K_w$ and hence $\#(S^m(w)) \geq 2$. Setting $m_0 = \max\{n, m_1\}$, we have the desired statement. $\qquad\qquad\square$

4.4   Construction of Adapted Metric II

In this section, we study a sufficient condition for the existence of an adapted metric to a given weight function under the basic framework presented in Sect. 4.3. This is a continuation of the study of Sect. 4.1.

As in the previous sections, $(T, \mathcal{A}, \phi)$ is a locally finite tree with the root ϕ, $(X, \mathcal{O})$ is a compact metrizable topological space with no isolated point, $K : T \to \mathcal{C}(X, \mathcal{O})$ is a minimal partition. We also assume that $\sup_{w \in T} \#(S(w)) < +\infty$. Moreover, we assume that $d \in \mathcal{D}_{A,e}(X, \mathcal{O})$ and the basic framework given in Sect. 4.3, i.e. the conditions (BF1) and (BF2) are satisfied.

In this section, we fix $g \in \mathcal{G}_e(T)$ satisfying $g \underset{\mathrm{GE}}{\sim} d$.

Remark The modification of T in the previous section does not affect the relation $\underset{\mathrm{GE}}{\sim}$ and the exponentiality of g.

Definition 4.4.1 Let $\varphi : T \to (0, \infty)$. For $M \geq 1$ and $w \in T$, define

$$\langle \varphi \rangle_M(w) = \min_{v \in \Gamma_M(w)} \varphi(v)$$

and

$$\Pi_M^\varphi(w) = \min_{v \in \Gamma_M(w)} \frac{\varphi(v)}{\varphi(\pi(v))}.$$

The next lemma, which holds without (BF1) and (BF2), gives a sufficient condition for a non-negative function on T to be balanced.

Lemma 4.4.2 *Let* $\varphi : T \to (0.\infty)$ *and let* $M_1, M_2 \in \mathbb{N}$. *Suppose that* (4.3.4) *holds for any* $v \in T$.

(1) *For any* $v \in T$,

$$\langle\varphi\rangle_{M_1+M_2}(v) \geq \Pi^{\varphi}_{M_1+M_2}(v) \max\{\langle\varphi\rangle_{M_1+M_2}(u) | u \in \Gamma_{M_1}(\pi(v))\}.$$

(2) *Assume that*

$$\sum_{i=1}^{m} \Pi^{\varphi}_{M_1+M_2}(w(i)) \geq 1 \tag{4.4.1}$$

for any $w \in T$ *and* $(w(1), \dots, w(m)) \in C^{M_1}_w$. *Then*

$$\sum_{i=1}^{m} \langle\varphi\rangle_{M_1+M_2}(w(i)) \geq \max_{i=1,\dots,m} \langle\varphi\rangle_{M_1+M_2}(\pi(w(i)))$$

for any $w \in T$ *and* $(w(1), \dots, w(m)) \in C^{M_1}_w$. *In particular,* $\langle\varphi\rangle_{M_1+M_2}$ *is* M_1-*balanced.*

Proof For simplicity, set $\varphi_* = \langle\varphi\rangle_{M_1+M_2}$.

(1) There exists $v' \in \Gamma_{M_1+M_2}(v)$ such that $\varphi_*(v) = \varphi(v')$. Let $u \in \Gamma_{M_1}(\pi(v))$. Since $\pi(v) \in \Gamma_{M_1}(u)$, Proposition 4.3.4 shows that $\pi(\Gamma_{M_1+M_2}(v)) \subseteq \Gamma_{M_1+M_2}(u)$. Therefore, $\pi(v') \in \Gamma_{M_1+M_*}(u)$. Hence

$$\varphi_*(v) = \varphi(v') = \frac{\varphi(v')}{\varphi(\pi(v'))}\varphi(\pi(v')) \geq \Pi^{\varphi}_{M_1+M_2}(v) \max\{\varphi_*(u) | u \in \Gamma_{M_1}(\pi(v))\}.$$

(2) Let $(w(1), \dots, w(m)) \in C^{M_1}_w$. Then by (1),

$$\sum_{i=1}^{m} \varphi_*(w(i)) \geq \sum_{i=1}^{m} \Pi^{\varphi}_{M_1+M_2}(w(i)) \max\{\varphi_*(u) | u \in \Gamma_{M_1}(\pi(w(i)))\}. \tag{4.4.2}$$

Let $j \in \{1, \dots, m\}$. Our goal is to show

$$\sum_{i=1}^{m} \varphi_*(w(i)) \geq \varphi_*(\pi(w(j))). \tag{4.4.3}$$

If $\pi(w(j)) \in \Gamma_{M_1}(\pi(w(i)))$ for any $i = 1, \ldots, m$, then (4.4.2) and (4.4.1) imply

$$\sum_{i=1}^{m} \varphi_*(w(i)) \geq \sum_{i=1}^{m} \Pi_{M_1+M_2}^{\varphi}(w(i))\varphi_*(\pi(w(j))) \geq \varphi_*(\pi(w(j))). \qquad (4.4.4)$$

Suppose that there exists $k \in \{1, \ldots, m\}$ such that $\pi(w(k)) \notin \Gamma_{M_1}(\pi(w(j)))$. If $k < j$, then set $k_* = \max\{l \,|\, \pi(w(l)) \notin \Gamma_{M_1}(\pi(w(j)))\}$. Since $w(j-1) \in \Gamma_{M_1}(w(j))$, we see that $\pi(w(j-1)) \in \Gamma_{M_1}(\pi(w(j)))$. Hence $k_* \leq j-2$. So $(w(k_*+1), \ldots, w(j-1)) \in C_{w(j)}^{M_1}$. Therefore, (4.4.1) shows

$$\sum_{i=k_*+1}^{j-1} \Pi_{M_1+M_1}^{\varphi}(w(i)) \geq 1.$$

Using (4.4.2), we have

$$\sum_{i=1}^{m} \varphi_*(w(i)) \geq \sum_{i=k_*+1}^{j-1} \Pi_{M_1+M_*}^{\varphi}(w(i)) \max_{u \in \Gamma_{M_1}(\pi(w(i)))} \varphi_*(u) \geq \varphi_*(\pi(w(j))).$$

If $k > j$, similar arguments show (4.4.3) as well. $\square$

Using the last lemma, we have a sufficient condition for the existence of a metric which is adapted to a given weight function.

Theorem 4.4.3 *Let $M_1, M_2 \in \mathbb{N}$. Suppose that (4.3.4) holds for any $v \in T$. Assume that $g \in \mathcal{G}_e(T)$ and $g \underset{GE}{\sim} d$. If*

$$\sum_{i=1}^{m} \Pi_{M_1+M_2}^{g}(w(i)) \geq 1 \qquad (4.4.5)$$

for any $w \in T$ and $(w(1), \ldots, w(m)) \in C_w^{M_1}$, then there exists a metric $\rho \in \mathcal{D}_{\mathcal{A},e}(X)$ which is M_1-adapted to g and quasisymmetric to d.

Proof By Lemma 4.4.2, $\langle g \rangle_{M_1+M_2}$ is M_1-balanced. Since $g \underset{GE}{\sim} d$, we have $g \underset{BL}{\sim} \langle g \rangle_{M_1+M_2}$. Therefore Theorem 4.1.3 shows that there exists a metric $\rho \in \mathcal{D}(X)$ such that ρ is M_1-adapted to g. Note that g is tight because $g \underset{GE}{\sim} d$ and d is tight. Therefore, Theorem 3.1.14 implies that $g \underset{BL}{\sim} \rho$ and hence ρ is exponential. Moreover, the fact that $\rho \underset{GE}{\sim} d$ implies $\rho \underset{QS}{\sim} d$ by Theorem 3.6.6. $\square$

To utilize Theorem 4.4.3, we need to find M_1 and M_2 satisfying (4.3.4);

$$\pi(\Gamma_{M_1+M_2}(v)) \subseteq \Gamma_{M_2}(\pi(v))$$

for any $v \in T$. Recall that d is M_*-adapted as in (BF1). Since the metric ρ obtained in the above theorem is quasisymmetric to d, Theorem 3.6.6 implies that ρ is M_*-adapted, and conversely, d is M_1-adapted as well. If $M_* = \min\{M|d \text{ is } M\text{-adapted.}\}$, then it follows $M_1 \geq M_*$. This requirement on M_1 can make it hard to find M_1 and M_2. Replacing π by π^k, however, we have the following fact.

Proposition 4.4.4 *For any $M \geq 1$, there exists $k_M \geq 1$ such that if $v \in T$ and $k \geq k_M$, then $\pi^k(\Gamma_{M+M_*}(v)) \subseteq \Gamma_{M_*}(\pi^k(v))$.*

Proof Since d is M_*-adapted, it is (M_*+M)-adapted as well. Therefore, there exist $c_1, c_2 > 0$ such that

$$U_{M_*+M}^{hr}(x, r^n) \subseteq B_d(x, c_1 r^n) \quad \text{and} \quad B_d(x, c_2 r^n) \subseteq U_{M_*}^{hr}(x, r^n)$$

for any $x \in X$ and $n \geq 0$. Set $k_M = \min\{k|c_1 r^k < c_2\}$. Let $k \geq k_M$. If $|v| \geq k$, then

$$U_{M_*+M}^{hr}(x, r^{|v|}) \subseteq B_d(x, c_1 r^{|v|}) \subseteq B_d(x, c_2 r^{|v|-k}) \subseteq U_{M_*}^{hr}(x, r^{|v|-k}).$$

If $x \in O_v$, then this implies that $K_u \subseteq \cup_{w \in \Gamma_{M_*}(\pi^k(v))} K_w$ for any $u \in \Gamma_{M_*+M}(v)$. By Lemma 4.3.6, it follows that $\pi^k(u) \in \Gamma_{M_*}(\pi^k(v))$. If $|v| < k$, then $\pi^k(v) = \phi$ and the desired statement is trivial. $\square$

Definition 4.4.5 Let $q \in \mathbb{N}$. Define $T^{(q)} = \cup_{m \geq 0}(T)_{mq}$. Define $\pi_q : T^{(q)} \to T^{(q)}$ by $\pi_q = \pi^q$, which is the q-th iteration of π. We consider $T^{(q)}$ as a tree with the reference point ϕ under the natural tree structure inherited from T. Then $(T^{(q)})_m = (T)_{mq}$. Moreover, set $K^{(q)} = K|_{T^{(q)}}$.

Note that a horizontal M-jpath of $K^{(q)}$ is a horizontal M-jpath of K. Similarly, $\Gamma_M(w, K^{(q)}) = \Gamma_M(w, K)$ and $U_M(w, K^{(q)}) = U_M(w, K)$ for any $w \in T^{(q)}$ and $J_{M,mq}^h(K) = J_{M,m}^h(K^{(q)})$.

Definition 4.4.6

(1) For $w \in T$, define

$$\mathcal{C}_{w,k}(N_1, N_2, N) = \{(w(1), \ldots, w(m))|(w(1), \ldots, w(m)) \text{ is}$$

$$\text{a horizontal } N\text{-jpath}, w(j) \in S^k(\Gamma_{N_2}(w)) \text{ for any}$$

$$j = 1, \ldots, m, \Gamma_N(w(1)) \cap S^k(\Gamma_{N_1}(w)) \neq \emptyset \text{ and}$$

$$\Gamma_N(w(m)) \backslash S^k(\Gamma_{N_2}(w)) \neq \emptyset\}.$$

(2) For a weight function $g \in \mathcal{G}(T^{(k)})$ on $T^{(k)}$, define

$$\Pi_M^{g,k}(w) = \min_{v \in \Gamma_M(w)} \frac{g(v)}{g(\pi^k(v))}$$

for any $w \in T^{(k)}$.

Note that $\mathcal{C}_w^M = \mathcal{C}_{w,1}(0, M, M)$.

Replacing T by $T^{(k)}$ and applying Theorem 4.4.3, we obtain the following corollary.

Corollary 4.4.7 *Let $M_1 \in \mathbb{N}$ and let $k \geq k_{M_1}$, where k_{M_1} is the constant appearing in Proposition 4.4.4 in case $M = M_1$. Assume that $g \in \mathcal{G}_e(T^{(k)})$ and $g \underset{GE}{\sim} d$ as weight functions on $T^{(k)}$. If*

$$\sum_{i=1}^{m} \Pi_{M_1+M_*}^{g,k}(w(i)) \geq 1$$

for any $w \in T^{(k)}$ and $(w(1), \ldots, w(m)) \in \mathcal{C}_{w,k}(0, M_1, M_1)$, then there exists a metric $\rho \in \mathcal{D}_{A,e}(X)$ such that ρ is M_*-adapted to g and ρ is quasisymmetric to d.

Remark It is easy to see that any weight function $g \in \mathcal{G}_e(T^{(k)})$ can be extended to a weight function $\widetilde{g} \in \mathcal{G}_e(T)$, i.e. there exists a weight function $\widetilde{g} \in \mathcal{G}_e(T)$ such that $\widetilde{g}|_{T^{(k)}} = g$. Since $\widetilde{g}$ is exponential, we can see that for any $M \geq 1$, there exist $c_1, c_2 > 0$ such that

$$c_1 D_M^g(x, y) \leq D_M^{\widetilde{g}}(x, y) \leq c_2 D_M^g(x, y)$$

for any $x, y \in X$. Therefore, the metric ρ obtained in the above corollary is M_*-adapted to $\widetilde{g}$ as well.

4.5 Construction of Ahlfors Regular Metric II

In this section, making use of Theorem 4.2.2 and Corollary 4.4.7, we are going to establish a sufficient condition for the existence of an adapted metric ρ and a measure μ where μ is Ahlfors regular with respect to the metric ρ.

As in the previous sections, $(T, \mathcal{A}, \phi)$ is a locally finite tree with the root ϕ, $(X, \mathcal{O})$ is a compact metrizable topological space with no isolated point, $K : T \to \mathcal{C}(X, \mathcal{O})$ is a minimal partition and $\sup_{w \in T} \#(S(w)) < +\infty$. Moreover, we keep employing the basic framework, i.e. the conditions (BF1) and (BF2).

Our main theorem of this section is as follows:

Theorem 4.5.1 *Let $M_1 \in \mathbb{N}$. Assume that $k \geq \max\{m_0, k_{M_1}, k_{M_*}\}$, where m_0 is the constant in Proposition 4.3.5 in case of $M = 1$ and k_{M_1} and k_{M_*} are the constants in Proposition 4.4.4 in case of $M = M_1$ and M_* respectively. If there exists $\varphi :$ $T^{(k)} \to [0, 1]$ such that*

$$\sum_{i=1}^{m} \varphi(w(i)) \geq 1$$

for any $w \in T^{(k)}$ and $(w(1), \ldots, w(m)) \in \mathcal{C}_{w,k}(0, M_1, M_1)$ and

$$\sum_{v \in S^k(w)} \varphi(v)^p < \frac{1}{4}(L_*)^{-2(M_1+2M_*)}, \tag{4.5.1}$$

for any $w \in T^{(k)}$, then there exist a family of metrics $\{\rho_t\}_{t \in [0,1]} \subseteq \mathcal{D}_{\mathcal{A},e}(X)$ and a family of Borel regular probability measures $\{\mu_t\}_{t \in [0,1]}$ on X such that

(a) $\rho_t \underset{QS}{\sim} d$ and μ_t is p-Ahlfors regular with respect to ρ_t for any $t \in [0, 1]$.
(b) ρ_t and ρ_s are not bi-Lipschitz equivalent if $t \neq s$.

Remark Examining the proof below, even if the constant $\frac{1}{4}$ in (4.5.1) is replaced by $\frac{1}{2}$, one can still prove the existence of a single metric $\rho \in \mathcal{D}_{\mathcal{A},e}(X)$ with $\rho \underset{QS}{\sim} d$ and a Borel regular measure μ which is p-Ahlfors regular with respect to ρ.

Remark By the choice of k in the above theorem, it follows that for any $v \in T$,

$$\pi^k(\Gamma_{M_1+M_*}(v)) \cup \pi^k(\Gamma_{2M_*}(v)) \subseteq \Gamma_{M_*}(\pi^k(v)).$$

The rest of this section is devoted to a proof of this theorem. First we present two key lemmas.

Lemma 4.5.2 *Let V be a countable set. Let $\{V(v)\}_{v \in V}$ be a family of finite subsets of V satisfying $v \in V(v)$ for any $v \in V$ and $v \in V(u)$ if and only if $u \in V(v)$ for any $u, v \in V$. Define*

$$V(A) = \bigcup_{v \in A} V(v)$$

for any $A \subseteq V$. Then for any $f : V \to [0, \infty)$, there exists $\sigma : V \to [0, \infty)$ such that

$$f(v) \leq \min\{\sigma(u) | u \in V(v)\} \leq \sigma(v) \leq \max_{u \in V(v)} f(u)$$

for any $v \in V$ and

$$\sum_{v \in U} \sigma(v)^p \leq \left(\max_{v \in V(U)} \#(V(v)) \right) \sum_{v \in V(U)} f(v)^p. \tag{4.5.2}$$

for any finite subset $U \subseteq V$ and $p > 0$.

Proof Define $\sigma(v) = \max\{f(u) | u \in V(v)\}$. Since $v \in V(u)$ if and only if $u \in V(v)$, it follows that $f(v) \leq \sigma(u)$ for any $u \in V(v)$. Moreover, we have $v \in V(v)$. Hence $f(v) \leq \min\{\sigma(u) | u \in V(v)\} \leq \sigma(v)$. Furthermore,

$$\sum_{v \in U} \sigma(v)^p \leq \sum_{v \in U} \sum_{u \in V(v)} f(u)^p$$

$$\leq \sum_{u \in V(U)} \sum_{v \in V(u)} f(u)^p = \sum_{u \in V(U)} \#(V(u)) f(u)^p.$$

Hence (4.5.2) holds. □

Lemma 4.5.3 *Let $k \geq k_{M_*}$. Let $\kappa_0 \in (0, 1)$ and let $f : T^{(k)} \backslash \{\phi\} \to [\kappa_0, 1)$. Then there exists $g : T^{(k)} \to (0, 1]$ such that*

$$g(u) \geq \kappa_0 g(v) \tag{4.5.3}$$

if $(u, v) \in J_{M_*}^h$,

$$f(v) \leq \frac{g(v)}{g(\pi^k(v))} \leq \max_{u \in \Gamma_{M_*}(v)} f(u) \tag{4.5.4}$$

for any $v \in T^{(k)} \backslash \{\phi\}$ and

$$\sum_{v \in S^k(w)} \left(\frac{g(v)}{g(\pi^k(v))} \right)^p \leq (L_*)^{2M_*} \max \left\{ \sum_{u \in S^k(w')} f(u)^p \middle| w' \in \Gamma_{M_*}(w) \right\}. \tag{4.5.5}$$

for any $p > 0$ and $w \in T^{(k)}$.

Proof First we are going to construct $g : \cup_{n \geq 0}(T)_{kn} \to (0, 1]$ satisfying (4.5.3) and (4.5.4) inductively. Set $g(\phi) = 1$ and $g(w) = f(w)$ for any $w \in (T)_k$. Then (4.5.3) and (4.5.4) are satisfied for any $w \in (T)_k$. Assume that there exists $g : \cup_{n=0}^m (T)_{kn} \to (0, 1]$ satisfying (4.5.3) and (4.5.4) up to the m-th level. Define

$$g_1(v) = f(v) g(\pi^k(v)),$$

$$g_2(v) = \kappa_0 \max_{u \in \Gamma_{M_*}(v)} g_1(u),$$

$$g(v) = \max\{g_1(v), g_2(v)\}.$$

for any $v \in (T)_{k(m+1)}$. First we are going to show (4.5.4) for $u \in (T)_{k(m+1)}$. If $g(v) = g_1(v)$, then $f(v) = \frac{g(v)}{g(\pi^k(v))} \leq \max_{u \in \Gamma_{M_*}(v)} f(u)$. Hence we have (4.5.4). If $g(v) = g_2(v)$, then $g_1(v) \leq g(v)$ and hence $f(v) \leq \frac{g(v)}{g(\pi^k(v))}$. There exists $u \in \Gamma_{M_*}(v)$ such hat $g(v) = \kappa_0 f(u) g(\pi^k(u))$. Since $(\pi^k(u), \pi^k(v)) \in J^h_{M_*}$, we have $g(\pi^k(u)) \leq (\kappa_0)^{-1} g(\pi^k(v))$ by (4.5.3) for $(T)_{km}$. Therefore,

$$\frac{g(v)}{g(\pi^k(v))} \leq f(u) \leq \max_{w \in \Gamma_{M_*}(v)} f(w).$$

Thus we have shown (4.5.4) for $(T)_{k(m+1)}$.

Next to show (4.5.3) for $(T)_{k(m+1)}$, we need the following fact.

Claim If $(u, v), (v, v') \in J^h_{M_*, k(m+1)}$, then $g_1(u) \geq (\kappa_0)^2 g_1(v')$.

Proof of Claim Note that $\pi^k(\Gamma_{2M_*}(u)) \subseteq \Gamma_{M_*}(\pi^k(u))$ because $k \geq k_{M_*}$. Hence $\pi^k(v') \in \Gamma_{M_*}(\pi^k(u))$ and so we have $(\pi^k(v'), \pi^k(u)) \in J^h_{M_*, km}$. Therefore, by (4.5.3), if $g_1(u) < (\kappa_0)^2 g_1(v')$, then

$$(\kappa_0)^2 g_1(v') > g_1(u) = f(u) g(\pi^k(u)) \geq \kappa_0 g(\pi^k(u)) \geq (\kappa_0)^2 g(\pi^k(v')),$$

and so $g_1(v') = f(v') g(\pi^k(v')) > g(\pi^k(v'))$. This contradicts the fact that $f(v') \leq 1$ and hence we have confirmed the claim.   □

Now let $(u, v) \in J^h_{M_*, k(m+1)}$. If $g(v) = g_2(v)$, then there exists $v' \in \Gamma_{M_*}(v)$ such that $g(v) = \kappa_0 g_1(v')$. By Claim, it follows that

$$g(u) \geq g_1(u) \geq (\kappa_0)^2 g_1(v') = \kappa_0 g(v).$$

If $g(v) = g_1(v)$, then $g(u) \geq g_2(u) \geq \kappa_0 g_1(v) = \kappa_0 g(v)$. Thus (4.5.3) holds for $(T)_{k(m+1)}$.

Using this construction of g inductively, we obtain $g : T \to (0, 1]$ satisfying (4.5.3) and (4.5.4) at every level. Next we are going to proof (4.5.5). Note that $\bigcup_{v \in S^k(w)} \Gamma_{M_*}(v) \subseteq \bigcup_{w' \in \Gamma_{M_*}(w)} S^k(w')$. By (4.5.4),

$$\sum_{v \in S^k(w)} \left(\frac{g(v)}{g(\pi^k(v))} \right)^p \leq \sum_{v \in S^k(w)} \sum_{u \in \Gamma_{M_*}(v)} f(u)^p$$

$$= \sum_{u \in \bigcup_{v \in S^k(w)} \Gamma_{M_*}(v)} \#(\Gamma_{M_*}(u) \cap S^k(w)) f(u)^p \leq (L_*)^{M_*}$$

$$\times \sum_{u \in \bigcup_{v \in S^k(w)} \Gamma_{M_*}(v)} f(u)^p$$

$$\leq (L_*)^{M_*} \sum_{w' \in \Gamma_{M_*}(w)} \sum_{u \in S^k(w')} f(u)^p$$

$$\leq (L_*)^{2M_*} \max \left\{ \sum_{u \in S^k(w')} f(u)^p \, \middle| \, w' \in \Gamma_{M_*}(w) \right\}.$$

□

Proof of Theorem 4.5.1 Set $\eta = \frac{1}{4}(L_*)^{-2(2M_*+M_1)}$. Assume that $\varphi : T^{(k)} \to [0, 1]$ satisfies

$$\sum_{i=1}^{m} \varphi(w(i)) \geq 1 \qquad (4.5.6)$$

for any $w \in T$ and $(w(1), \ldots, w(m)) \in \mathcal{C}_{w,k}(0, M_1, M_1)$ and

$$\sum_{v \in S^k(w)} \varphi(v)^p < \eta$$

for any $v \in T^{(k)}$. Define

$$\widetilde{\varphi}(v) = \left(\varphi(v)^p + \frac{\eta}{(N_*)^k} \right)^{1/p}$$

for any $v \in T^{(k)}$. Then (4.5.6) still holds if we replace φ by $\widetilde{\varphi}$. Moreover,

$$\sum_{v \in S^k(w)} \widetilde{\varphi}(v)^p < 2\eta$$

and

$$\left(\frac{\eta}{(N_*)^k} \right)^{1/p} \leq \widetilde{\varphi}(v) \leq \left(\frac{1 + (N_*)^k}{(N_*)^k} \eta \right)^{1/p}.$$

Set $M_3 = M_1 + M_*$. Letting $V = (T)_{km}$, $V(v) = \Gamma_{M_3}(v)$ and $f = \widetilde{\varphi}$ and applying Lemma 4.5.2, we obtain $\psi : T^{(k)} \to [0, \infty)$ satisfying

$$\widetilde{\varphi}(v) \leq \langle \psi \rangle_{M_3}(v) \leq \psi(v) \leq \left(\frac{1 + (N_*)^k}{(N_*)^k} \eta \right)^{1/p}$$

for any $v \in T^{(k)}$ and

$$\sum_{v \in S^k(w)} \psi(v)^p \le (L_*)^{M_3} \sum_{u \in \cup_{v \in S^k(w)} \Gamma_{M_3}(v)} \widetilde{\varphi}(u)^p$$

$$\le (L_*)^{M_3} \sum_{w' \in \Gamma_{M_3}(w)} \sum_{u \in S^k(w')} \widetilde{\varphi}(u)^p < 2(L_*)^{2M_3} \eta.$$

Next step is to use Lemma 4.5.3. Set $\kappa_0 = \left(\frac{\eta}{(N_*)^k}\right)^{1/p}$ and $\kappa_1 = 2^{-1/p}$. Applying Lemma 4.5.3 with $f = \psi$, we obtain $g : T^{(k)} \to (0, 1]$ satisfying (4.5.3)–(4.5.5). Define $\tau(w) = g(w)/g(\pi^k(w))$ for any $w \in T^{(k)} \backslash \{\phi\}$. Then by (4.5.5)

$$\tau(v)^p \le \sum_{u \in S^k(w)} \tau(u)^p < \frac{1}{2}$$

for any $w \in T^{(k)}$ and $v \in S^k(w)$ and by (4.5.4), for any $w \in T^{(k)} \backslash \{\phi\}$,

$$\kappa_0 \le \psi(w) \le \tau(w) \le \kappa_1.$$

To construct desired weight function, we need to modify τ once more. Since $k \ge m_0$, Proposition 4.3.5 shows that, for any $w \in T^{(k)}$, there exists $v_w \in S^k(w)$ such that $\Gamma_1(v_w) \subseteq S^k(w)$. Furthermore, $\#(S^k(w)) \ge 2$ by Proposition 4.3.5. Let $\epsilon \in [0, 2^{1/p} - 1)$. Define

$$\sigma_\epsilon(v) = \begin{cases} (1 + \epsilon)\tau(v) & \text{if } v \ne v_{\pi^k(v)}, \\ \left(1 - (1 + \epsilon)^p \sum_{u \in S^k(\pi^k(v)) \backslash \{v\}} \tau(u)^p\right)^{1/p} & \text{if } v = v_{\pi^k(v)}. \end{cases}$$

For ease of notation, we use σ instead of σ_ϵ for a while. If $v \ne v_{\pi^k(v)}$, then

$$\sigma(v)^p = (1 + \epsilon)^p \tau(v)^p \le (1 + \epsilon)^p \sum_{u \in S^k(\pi^k(v))} \le \frac{1}{2}(1 + \epsilon)^p < 1.$$

If $v = v_{\pi^k(v)}$, then

$$\sigma(v)^p = 1 - (1 + \epsilon)^p \sum_{u \in S^k(\pi^k(v))} \tau(u)^p + (1 + \epsilon)^p \tau(v)^p \ge (1 + \epsilon)^p \tau(v)^p$$

Hence

$$\kappa_0 \le \tau(v) \le \sigma(v) \le \max\{(1 - (\kappa_0)^p)^{1/p}, \kappa_1(1 + \epsilon)\} < 1 \tag{4.5.7}$$

and

$$\sum_{v \in S^k(w)} \sigma(v)^p = 1 \qquad (4.5.8)$$

for any $w \in T^{(k)}$. Since

$$\widetilde{\varphi}(v) \le \langle \psi \rangle_{M_3}(v) \quad \text{and} \quad \psi(v) \le \tau(v) \le \sigma(v),$$

it follows that $\widetilde{\varphi}(v) \le \langle \sigma \rangle_{M_3}(v)$. Hence

$$\sum_{i=1}^{m} \langle \sigma \rangle_{M_3}(w(i)) \ge 1 \qquad (4.5.9)$$

for any $w \in T^{(k)}$ and $(w(1), \ldots, w(m)) \in \mathcal{C}_{w,k}(0, M_1, M_1)$. Define $\widetilde{g}_\epsilon(w)$ inductively by $\widetilde{g}_\epsilon(\phi) = 1$ and

$$\widetilde{g}_\epsilon(w) = \sigma_\epsilon(w)\widetilde{g}_\epsilon(\pi^k(w)).$$

Suppose that $u, v \in (T)_{kn}$, $u \ne v$ and $u \in \Gamma_1(v)$. Then $\pi^{kl}(u) \ne \pi^{kl}(v)$ and $\pi^{k(l+1)}(u) = \pi^{k(l+1)}(v)$ for some $l \ge 0$. Note that $\pi^{kj}(u) \in T_{\pi^{kl}(u)}$, $\pi^{kj}(v) \in T_{\pi^{kl}(v)}$ and $\pi^{kj}(u) \in \Gamma_1(\pi^{kj}(v))$ for any $j = 0, \ldots, l-1$. Hence we see that

$$\frac{\widetilde{g}_\epsilon(u)}{\widetilde{g}_\epsilon(v)} = \frac{\sigma(u)\sigma(\pi^k(u)) \cdots \sigma(\pi^{kl}(u))}{\sigma(v)\sigma(\pi^k(v)) \cdots \sigma(\pi^{kl}(v))} = \frac{(1+\epsilon)^l \tau(u) \cdots \tau(\pi^{k(l-1)}(u))\sigma(\pi^{kl}(u))}{(1+\epsilon)^l \tau(v) \cdots \tau(\pi^{k(l-1)}(v))\sigma(\pi^{kl}(v))}.$$

On the other hand,

$$\frac{g(u)}{g(v)} = \frac{\tau(u) \cdots \tau(\pi^{k(l-1)}(u))\tau(\pi^{kl}(u))}{\tau(v) \cdots \tau(\pi^{k(l-1)}(v))\tau(\pi^{kl}(v))}.$$

Thus if $\kappa_2 = \max\{\kappa_1, \left(1 - (\kappa_0)^p\right)^{1/p}, 2^{-1/p}(1+\epsilon)\}$, then

$$\frac{\widetilde{g}_\epsilon(u)}{\widetilde{g}_\epsilon(v)} = \frac{g(u)}{g(v)} \frac{\sigma(\pi^{kl}(u))}{\sigma(\pi^{kl}(v))} \frac{\tau(\pi^{kl}(v))}{\tau(\pi^{kl}(u))} \ge \kappa_0 \frac{\kappa_0}{\kappa_2} \frac{\kappa_0}{\kappa_1}. \qquad (4.5.10)$$

By (4.5.7), the weight function $\widetilde{g}_\epsilon$ on $T^{(k)}$ is exponential. By (4.5.10), $\widetilde{g}_\epsilon \underset{\text{GE}}{\sim} d$ as weight functions on $T^{(k)}$. These facts with (4.5.9) enable us to apply Corollary 4.4.7 and deduce that there exists a metric $\rho_\epsilon \in \mathcal{D}_{\mathcal{A},e}(X)$ such that ρ_ϵ is M_*-adapted to $\widetilde{g}_\epsilon$, $\rho_\epsilon \underset{\text{BL}}{\sim} \widetilde{g}_\epsilon$ and $\rho_\epsilon \underset{\text{QS}}{\sim} d$. Moreover, by (4.5.8), $\widetilde{g}_\epsilon$ satisfies the $T^{(k)}$-version of (4.2.2). Hence applying Theorem 4.2.2 to $\widetilde{g}_\epsilon$ on $T^{(k)}$, we verify the existence of a Borel regular probability measure μ_ϵ which is p-Ahlfors regular with respect to

ρ_ϵ. Let $\epsilon_1 \neq \epsilon_2$. There exists $\omega = (\phi, w(1), w(2), \ldots) \in \Sigma$ such that $w(m) \neq v_{w(m-1)}$ for any $m \geq 1$. Then $\widetilde{g}_{\epsilon_1}(w(nk)) = (1 + \epsilon_1)^n \tau(w(k)) \cdots \tau(w(nk))$ and $\widetilde{g}_{\epsilon_2} = (1 + \epsilon_2)^n \tau(w(1)) \cdots \tau(w(nk))$. This shows that $\widetilde{g}_{\epsilon_1} \underset{\mathrm{BL}}{\not\sim} \widetilde{g}_{\epsilon_2}$ and hence ρ_{ϵ_1} and ρ_{ϵ_2} are not bi-Lipschitz equivalent. □

4.6 Critical Index of p-Energies and the Ahlfors Regular Conformal Dimension

Finally in this section, we establish the characterization of the Ahlfors regular conformal dimension as the critical index of p-energies.

As in the previous sections, $(T, \mathcal{A}, \phi)$ is a locally finite tree with the root ϕ, $(X, \mathcal{O})$ is a compact metrizable topological space with no isolated point, $K : T \to \mathcal{C}(X, \mathcal{O})$ is a minimal partition. We also assume that $\sup_{w \in T} \#(S(w)) < +\infty$.

Throughout this and the following sections, we fix $d \in \mathcal{D}_{A,\epsilon}(X, \mathcal{O})$ satisfying the basic framework, i.e. (BF1) and (BF2) in Sect. 4.3.

First we recall the definition of Ahlfors regular conformal dimension.

Definition 4.6.1 Let (X, d) be a metric space. The Ahlfors regular conformal dimension, AR conformal dimension for short, of a metric space (X, d) is defined as

$$\dim_{AR}(X, d) = \inf\{\alpha | \text{there exist a metric } \rho \text{ on } X \text{ and a Borel regular measure } \mu \text{ on}$$

$$X \text{ such that } \rho \underset{\mathrm{QS}}{\sim} d \text{ and } \mu \text{ is } \alpha\text{-Ahlfors regular with respect to } \rho\}.$$

The definition of p-energy $\mathcal{E}_p(f | V, E)$ of a function f on a graph (V, E) is as follows.

Definition 4.6.2 Let $G = (V, E)$ be a (non-directed) graph. For $f : V \to \mathbb{R}$ and $p > 0$, define

$$\mathcal{E}_p(f | V, E) = \frac{1}{2} \sum_{(x,y) \in E} |f(x) - f(y)|^p.$$

If $E = \emptyset$, then we define $\mathcal{E}_p(f, V, E) = 0$ for any $f : V \to \mathbb{R}$. Let $V_1, V_2 \subseteq V$. Assume that $V_1 \cap V_2 = \emptyset$. Define

$$\mathcal{F}_F(V, V_1, V_2) = \{f | f : V \to [0, \infty), f|_{V_1} \geq 1, f|_{V_2} \equiv 0\}$$

and

$$\mathcal{E}_p(V, E, V_1, V_2) = \inf\{\mathcal{E}_p(f | V, E) | f \in \mathcal{F}_F(V, V_1, V_2)\}$$

For $p = 2$, on the analogy of electric circuits, the quantity $\mathcal{E}_2(V, E, V_1, V_2)$ is considered as the conductance (and its reciprocal is considered as the resistance) between V_1 and V_2. In the same way, we may regard $\mathcal{E}_p(V, E, V_1, V_2)$ as the p-conductance between V_1 and V_2.

Applying the above definition to the horizontal graphs $((T)_m, J^h_{N,m})$, we define the critical index $I_{\mathcal{E}}(N_1, N_2, N)$ of p-energies.

Definition 4.6.3 Let N_1, N_2 and N be integers satisfying $N_1 \geq 0$, $N_2 > N_1$ and $N \geq 1$. For $p > 0$, define

$$\mathcal{E}_{p,k}(N_1, N_2, N) = \sup_{w \in T} \mathcal{E}_p((T)_{|w|+k}, J^h_{N,|w|+k}, S^k(\Gamma_{N_1}(w)), S^k(\Gamma_{N_2}(w))^c),$$

$$\overline{\mathcal{E}}_p(N_1, N_2, N) = \limsup_{k \to \infty} \mathcal{E}_{p,k}(N_1, N_2, N)$$

and

$$\underline{\mathcal{E}}_p(N_1, N_2, N) = \liminf_{k \to \infty} \mathcal{E}_{p,k}(N_1, N_2, N).$$

Furthermore, define

$$I_{\mathcal{E}}(N_1, N_2, N) = \inf\{p | \underline{\mathcal{E}}_p(N_1, N_2, N) = 0\}.$$

The last quantity $I_{\mathcal{E}}(N_1, N_2, N)$ is called the critical index of p-energies. Two values $\overline{\mathcal{E}}_p(N_1, N_2, N)$ and $\underline{\mathcal{E}}_p(N_1, N_2, N)$ represents the asymptotic behavior of the p-conductance between $\Gamma_{N_1}(w)$ and the complement of $\Gamma_{N_2}(w)$ as we refine the graph between those two sets.

Remark Ordinary we don't need $\mathcal{E}_p$ for $p = 0$ in the course of the discussions in this monograph. For the sake of completeness, we may, however, define

$$\mathcal{E}_{0,k}(N_1, N_2, N) = \sup_{w \in T} \#(S^k(\Gamma_{N_2}(w))).$$

Theorem 4.6.4 *For any $N \geq 1$,*

$$I_{\mathcal{E}}(N_1, N_2, N) = \dim_{AR}(X, d)$$

if $N_1 + M_* \leq N_2$.

Remark As is shown in Theorem 4.6.9, even if we replace $\underline{\mathcal{E}}_p$ by $\overline{\mathcal{E}}_p$ in the definition of $I_{\mathcal{E}}(N_1, N_2, N)$, the value of $I_{\mathcal{E}}$ is the same.

Up to now we have considered the critical exponent for p-energies associated with the graphs $\{((T)_m, J^h_{N,m})\}_{m \geq 0}$. In fact, the critical exponent is robust with respect to certain modifications of graphs as one can see in Theorem 4.6.9. The admissible class of modified graphs is called a proper system of horizontal networks.

Definition 4.6.5 A sequence of graphs $\{(\Omega_m, E_m)\}_{m \geq 0}$ is called a proper system of horizontal networks with indices (N, L_0, L_1, L_2) if and only if the following conditions (N1), (N2), (N3), (N4) and (N5) are satisfied:

(N1)   For every $m \geq 0$, $\Omega_m = A_m \cup V_m$ where $A_m \subseteq (T)_m$ and $V_m \subseteq X$.

(N2)   For any $m \geq 0$ and $w \in (T)_m$, $\Omega_{m,w} \neq \emptyset$, where $\Omega_{m,w}$ is defined as

$$\Omega_{m,w} = (\{w\} \cap A_m) \cup (V_m \cap K_w).$$

(N3)   If we define

$$E_m(u, v) = \{(x, y) | (x, y) \in E_m, x \in \Omega_{m,u}, y \in \Omega_{m,v}\},$$

then

$$\#(E_m(u, v)) \leq L_0$$

for any $m \geq 0$ and $u, v \in (T)_m$

(N4)   For any $(x, y) \in E_m$, $x \in \Omega_{m,u}$ and $y \in \Omega_{m,v}$ for some $(u, v) \in J^h_{N,m}$.

(N5)   For any $m \geq 0$, $(u, v) \in J^h_{L_1,m}$, $x \in \Omega_{m,u}$ and $y \in \Omega_{m,v}$, there exist $(x_1, \ldots, x_n)$ and $(w(1), \ldots, w(n))$ such that $w(i) \in \Gamma_{L_2}(u)$ for any $i \geq 1, \ldots, n$, $(x_i, x_{i+1}) \in E_m(w(i), w(i+1))$ for any $i = 1, \ldots, n-1$ and $x_1 = x, x_n = y, w(1) = u, w(n) = v$.

Example 4.6.6 Let $\Omega_*^{(N)} = \{((T)_m, J^h_{N,m})\}_{m \geq 0}$. Then $\Omega_*^{(N)}$ is a proper system of horizontal networks with indices $(N, 1, 1, 1)$.

Example 4.6.7 (The Sierpinski Carpet; Fig. 4.1) Let d be the Euclidean metric (divided by $\sqrt{2}$ so that the diameter of $[0, 1]^2$ is one.) Then $h_{1/3} \underset{\mathrm{BL}}{\sim} d$. Obviously, d is 1-adapted to the weight function $h_{1/3}$, exponential and uniformly finite. In this

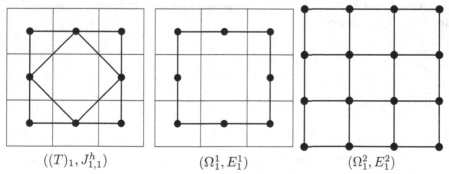

$$((T)_1, J^h_{1,1}) \qquad\qquad (\Omega_1^1, E_1^1) \qquad\qquad (\Omega_1^2, E_1^2)$$

$\bullet$'s are vertices and thick lines are edges.

Fig. 4.1 Proper systems of horizontal networks: the Sierpinski carpet

case, the original edges of the horizontal graph $((T)_m, J^h_{1,m})$ contain slanting edges, which are $(w, v) \in (T)_m \times (T_m)$ with $K_w \cap K_v$ being a single point. Even if all the slanting edges are deleted, we still have a proper system of horizontal networks $\{(\Omega^1_m, E^1_m)\}_{m \geq 0}$ given by

$$\Omega^1_m = (T)_m$$

and

$$E^1_m = \{(w, v) | w, v \in (T)_m, \dim_H(K_w \cap K_v, d) = 1\}.$$

The indices of $\{(\Omega^1_m, E^1_m)\}_{m \geq 0}$ is $(1, 1, 1, 2)$.

There is another natural proper system of horizontal networks. Note that the four points $p_1 = (0, 0)$, $p_3 = (1, 0)$, $p_5 = (1, 1)$ and $p_7 = (0, 1)$ are the corners of the square $[0, 1]^2$. Define $\Omega^2_0 = \{p_1, p_3, p_5, p_7\}$ and

$$E^2_0 = \{(p_i, p_j) | \text{the line segment } p_i p_j \text{ is one of the four line segments}$$

$$\text{of the boundary of } [0, 1]^2\}.$$

For $m \geq 1$, we define

$$\Omega^2_m = \cup_{w \in (T)_m} F_w(\Omega^2_0)$$

and

$$E^2_m = \{(F_w(p_j), F_w(p_j)) | w \in (T)_m, (p_i, p_j) \in E^2_0\}.$$

Then $\{(\Omega^2_m, E^2_m)\}_{m \geq 0}$ is a proper system of horizontal networks with indices $(1, 6, 1, 1)$. In this case all the vertices are the points in the Sierpinski carpet, and the length between the end points of an edge in E^2_m is 3^{-m}.

Notation Let $\Omega = \{(\Omega_m, E_m)\}_{m \geq 0}$ be a proper system of horizontal networks. For any $U \subseteq (T)_m$, we define

$$\Omega_m(U) = \bigcup_{v \in U} \Omega_{m,v}. \tag{4.6.1}$$

Furthermore, for $w \in T, k \geq 0$ and $n \geq 0$, we define

$$\Omega^k(w, n) = \Omega_{|w|+k}(S^k(\Gamma_n(w))) \tag{4.6.2}$$

$$\Omega^{k,c}(w, n) = \Omega_{|w|+k}((T)_{|w|+k} \setminus S^k(\Gamma_n(w))). \tag{4.6.3}$$

In the same manner as the original case, we define the p-conductances and the critical index of p-energies for a proper system of horizontal networks as follows.

Definition 4.6.8 Let $\Omega = \{(\Omega_m, E_m)\}_{m \geq 0}$ be a proper system of horizontal networks. For $p > 0$, define

$$\mathcal{E}_{p,k,w}(N_1, N_2, \Omega) = \mathcal{E}_p(\Omega_{|w|+k}, E_{|w|+k}, \Omega^k(w, N_1), \Omega^{k,c}(w, N_2)),$$

$$\mathcal{E}_{p,k}(N_1, N_2, \Omega) = \sup_{w \in T} \mathcal{E}_{p,k,w}(N_1, N_2, \Omega),$$

$$\overline{\mathcal{E}}_p(N_1, N_2, \Omega) = \limsup_{k \to \infty} \mathcal{E}_{p,k}(N_1, N_2, \Omega),$$

$$\underline{\mathcal{E}}_p(N_1, N_2, \Omega) = \liminf_{k \to \infty} \mathcal{E}_{p,k}(N_1, N_2, \Omega),$$

$$\overline{I}_{\mathcal{E}}(N_1, N_2, \Omega) = \inf\{p | \overline{\mathcal{E}}_p(N_1, N_2, \Omega) = 0\},$$

$$\underline{I}_{\mathcal{E}}(N_1, N_2, \Omega) = \inf\{p | \underline{\mathcal{E}}_p(N_1, N_2, \Omega) = 0\}.$$

Comparing Definitions 4.6.3 and 4.6.8, we notice that

$$\mathcal{E}_{p,k}(N_1, N_2, N) = \mathcal{E}_{p,k}(N_1, N_2, \Omega_*^{(N)}).$$

Thus Theorem 4.6.4 is a corollary of the following theorem.

Theorem 4.6.9 *Let Ω be a proper system of horizontal networks. If $N_2 \geq N_1 + M_*$, then*

$$\overline{I}_{\mathcal{E}}(N_1, N_2, \Omega) = \underline{I}_{\mathcal{E}}(N_1, N_2, \Omega) = \dim_{AR}(X, d).$$

The following corollary will be concluded from the proof of the above theorem.

Corollary 4.6.10 *If $p > \dim_{AR}(X, d)$, then there exists a family of metrics $\{\rho_t\}_{t \in [0,1]} \subseteq \mathcal{D}_{A,e}(X)$ such that $\rho_t \underset{QS}{\sim} d$ and ρ_t is p-Ahlfors regular for any $t \in [0, 1]$, and if $t \neq s$, then ρ_t and ρ_s are not bi-Lipschitz equivalent.*

Before a proof of this theorem, we are going to present a corollary which ensures the finiteness of $\dim_{AR}(X, d)$. To begin with, we need to define growth rates of volumes.

Definition 4.6.11 Define

$$\overline{N}_* = \limsup_{n \to \infty} \left(\sup_{w \in T} \#(S^n(\Gamma_{N_2}(w))) \right)^{\frac{1}{n}}$$

$$\underline{N}_* = \liminf_{n \to \infty} \left(\sup_{w \in T} \#(S^n(\Gamma_{N_2}(w))) \right)^{\frac{1}{n}}.$$

It is easy to see that

$$\underline{N}_* \leq \overline{N}_* \leq N_*.$$

The quantities $\overline{N}_*$ and $\underline{N}_*$ appear to depend on the value of N_2 but they do not as is shown in the next lemma.

Lemma 4.6.12

$$\overline{N}_* = \limsup_{n \to \infty} \left(\sup_{w \in T} \#(S^n(w)) \right)^{\frac{1}{n}}$$

$$\underline{N}_* = \liminf_{n \to \infty} \left(\sup_{w \in T} \#(S^n(w)) \right)^{\frac{1}{n}}.$$

Proof Since $S^n(w) \subseteq S^n(\Gamma_{N_2}(w))$, we have

$$\sup_{w \in T} \#(S^n(w)) \leq \sup_{w \in T} \#(S^n(\Gamma_{N_2}(w))).$$

On the other hand, by the fact that $\#(S^n(w)) \leq (N_*)^n$, there exists $w(n) \in T$ such that $\#(S^n(w(n)))$ attains the supremum. Note that

$$S^n(\Gamma_{N_2}(w)) = \bigcup_{v \in \Gamma_{N_2}(w)} S^n(v).$$

Therefore,

$$\#(S^n(\Gamma_{N_2}(w))) \leq \#(\Gamma_{N_2}(w))\#(S^n(w(n))) \leq (L_*)^{N_2} \sup_{w \in T} \#(S^n(w)).$$

$\square$

Corollary 4.6.13 *Let Ω be a proper system of horizontal networks. Then*

$$\dim_{AR}(X, d) \leq -\frac{\log \underline{N}_*}{\log r} \leq -\frac{\log N_*}{\log r},$$

where r is given in (BF2).

Now we start proving the theorem and the corollary.
Using the condition (N5), one can easily obtain the first lemma.

Lemma 4.6.14 *Let $\Omega = \{(\Omega_m, E_m)\}_{m \geq 0}$ be a proper system of horizontal networks with indices (N, L_0, L_1, L_2). Then Ω is a proper system of horizontal networks with indices $(N, L_0, nL_1, (n-1)L_1 + L_2)$ for any $n \geq 1$.*

Lemma 4.6.15 *Let $\Omega = \{(\Omega_m, E_m)\}_{m \geq 0}$ be a proper system of horizontal networks. Assume $N_2 \geq N_1 + M_*$. If ρ is a metric on X with $\mathrm{diam}(X, \rho) = 1$ and*

$\rho \underset{QS}{\sim} d$. Then for any $p > 0$, there exists $c > 0$ such that

$$\mathcal{E}_{p,k,w}(N_1, N_2, \Omega) \le c \sum_{u \in S^k(\Gamma_{N_2}(w))} \left(\frac{\rho(u)}{\rho(w)} \right)^p$$

for any $w \in T$ and $k \ge 0$.

Proof Since $\rho \underset{QS}{\sim} d$, Theorem 3.6.6 implies that $\rho \in \mathcal{D}_{A,e}(X)$ and $d \underset{GE}{\sim} \rho$. Moreover, ρ is M_*-adapted. Set $\rho(A, B) = \inf_{x \in A, y \in B} \rho(x, y)$ for $A, B \subseteq X$. For $w \in T$, define

$$f_w(x) = \begin{cases} \min\left\{ \dfrac{\rho(K_x, U_{N_2}(w)^c)}{\rho(U_{N_1}(w), U_{N_2}(w)^c)}, 1 \right\} & \text{if } x \in A_m, \\[3mm] \min\left\{ \dfrac{\rho(x, U_{N_2}(w)^c)}{\rho(U_{N_1}(w), U_{N_2}(w)^c)}, 1 \right\} & \text{if } x \in V_m. \end{cases}$$

for any $x \in \cup_{k \ge 0} \Omega_{|w|+k}$. If is easy to see that $f_w(x) = 1$ for any $x \in \Omega^k(w, N_1)$ and $f_w(x) = 0$ for any $x \in \Omega^{k,c}(w, N_2)$. Let $v \in \Gamma_{N_1}(w)$ and let $x \in O_v$. If $(w(1), \ldots, w(M_*+1))$ is a horizontal chain in $(T)_{|w|}$ and $x \in K_{w(1)}$, then $w(1) = v$ and $w(M_* + 1) \in \Gamma_{N_1+N_*}(w) \subseteq \Gamma_{N_2}(w)$. Hence Lemma 3.5.8 implies

$$U_{N_2}(w) \supseteq U_{M_*}^d(x, r^{|w|}) \supseteq U_{M_*}^\rho(x, \gamma\rho(w)) \supseteq B_\rho(x, \gamma'\rho(w)).$$

Thus $\rho(x, y) \ge \gamma'\rho(w)$ for any $x \in \cup_{v \in \Gamma_{N_1}(w)} O_v$ and $y \in U_{N_2}(w)^c$. Consequently

$$\rho(U_{N_1}(w), (U_{N_2}(w))^c) \ge \gamma'\rho(w). \tag{4.6.4}$$

Let (N, L_0, L_1, L_2) be the indices of Ω. Since $\rho \underset{GE}{\sim} d$, there exists $\kappa \in (0, 1)$ such that

$$\kappa\rho(u) \le \rho(v) \tag{4.6.5}$$

if $|u| = |v|$ and $K_u \cap K_v \ne \emptyset$. Note that if $A, B, C \subseteq X$, then

$$|\rho(A, B) - \rho(C, B)| \le \sup_{a \in A, c \in C} \rho(x, y). \tag{4.6.6}$$

For any $(u, v) \in J^h_{N, |w|+k}$ and $(x, y) \in E_{|w|+k}(u, v)$, if $(w(1), \ldots, w(N + 1))$ is a horizontal N-chain between u and v, then by (4.6.4)–(4.6.6),

$$|f_w(x) - f_w(y)| \le \frac{\sup_{a \in K_v, b \in K_w} \rho(a, b)}{\gamma' \rho(w)}$$

$$\le \frac{1}{\gamma' \rho(w)} \sum_{i=1}^{N+1} \rho(w(i)) \le (N + 1)\kappa^{-N}(\gamma')^{-1} \frac{\rho(u)}{\rho(w)}.$$

Hence

$$\sum_{(x,y) \in E_{|w|+k}(u,v)} |f_w(x) - f_w(y)|^p \le L_0((N + 1)\kappa^{-N}(\gamma')^{-1})^p \left(\frac{\rho(u)}{\rho(w)} \right)^p.$$

Set $c_1 = L_0((N + 1)\kappa^{-N}(\gamma')^{-1})^p$. Then the above inequality and the condition (N4) imply that

$$\mathcal{E}_{p,k,w}(N_1, N_2, \Omega) \le \frac{1}{2} \sum_{u \in S^k(\Gamma_{N_2}(w))} \sum_{v \in \Gamma_N(u)} \sum_{(x,y) \in E_{|w|+k}(u,v)} |f_w(x) - f_w(y)|^p$$

$$\le c_1(L_*)^N \sum_{u \in S^k(\Gamma_{N_2}(w))} \left(\frac{\rho(u)}{\rho(w)} \right)^p.$$

□

The next lemma yields the fact that $I_\mathcal{E}(N_1, N_2, \Omega) \le \dim_{AR}(X, d)$ if $N_2 \ge N_1 + M_*$.

Lemma 4.6.16 *Let $\Omega = \{(\Omega_m, E_m)\}_{m \ge 0}$ be a proper system of horizontal networks. Assume $N_2 \ge N_1 + M_*$. If $\dim_{AR}(X, d) < p$, then $\overline{\mathcal{E}}_p(N_1, N_2, \Omega) = 0$.*

Proof Since $\dim_{AR}(X, d) < p$, there exist $q \in [\dim_{AR}(X, d), p)$, a metric ρ, a Borel regular measure μ, and constants $c_1, c_2 > 0$ such that $d \underset{QS}{\sim} \rho$ and

$$c_1 s^q \le \mu(B_\rho(x, r)) \le c_2 s^q$$

for any $x \in X$ and $s \in [0, \text{diam}(X, \rho)]$. By Lemma 4.6.15,

$$\mathcal{E}_{p,k,w}(N_1, N_2, \Omega) \le c \sum_{u \in S^k(\Gamma_{N_2}(w))} \left(\frac{\rho(u)}{\rho(w)} \right)^p$$

$$\le c \max_{u \in S^k(\Gamma_{N_2}(w))} \left(\frac{\rho(u)}{\rho(w)} \right)^{p-q} \sum_{u \in S^k(\Gamma_{N_2}(w))} \left(\frac{\rho(u)}{\rho(w)} \right)^q.$$

$$(4.6.7)$$

Since ρ is exponential, there exist $\lambda \in (0, 1)$ and $c > 0$ such that $\rho(v) \leq c\lambda^k \rho(\pi^k(v))$ for any $v \in T$. Choose κ as in (4.6.5). If $u \in S^k(\Gamma_{N_2}(w))$, then

$$\rho(u) \leq c\lambda^k \rho(\pi^k(u)) \leq c\lambda^k \kappa^{-N_2} \rho(w). \tag{4.6.8}$$

On the other hand, by Theorem 4.2.2, we have (4.2.2) with $g = g_\rho$. Hence

$$\sum_{u \in S^k(v)} \rho(u)^q \leq c\rho(v)^q \leq c\kappa^{-N_2 q} \rho(w)^q$$

for any $v \in \Gamma_{N_2}(w)$. Thus

$$\sum_{u \in S^k(\Gamma_{N_2}(w))} \left(\frac{\rho(u)}{\rho(w)}\right)^q = \sum_{v \in \Gamma_{N_2}(w)} \sum_{u \in S^k(v)} \left(\frac{\rho(u)}{\rho(w)}\right)^q$$

$$\leq \#(\Gamma_{N_2}(w))c\kappa^{-N_2 q} \leq (L_*)^{N_2+1} c\kappa^{-N_2 q}. \tag{4.6.9}$$

Combining (4.6.7)–(4.6.9), we obtain

$$\mathcal{E}_{p,k,w}(N_1, N_2, \Omega) \leq c'\lambda^{(p-q)k}, \tag{4.6.10}$$

where c' is independent of w. Therefore, we conclude that $\overline{\mathcal{E}}_p(N_1, N_2, \Omega) = 0$. $\square$

The following version of discrete Hölder inequality will be used several times. It is obtained by applying the ordinary Hölder inequality to a sum of Dirac measures.

Lemma 4.6.17 *Let* $C_h(p, n) = \max\{n^{p-1}, 1\}$. *For any* $a_1, \ldots, a_n \in \mathbb{R}$,

$$\left(\sum_{i=1}^n |a_i|\right)^p \leq C_h(p, n) \sum_{i=1}^n |a_i|^p.$$

The following lemma enables us to apply Theorem 4.5.1 and to construct desired pair of a metric and a measure with Ahlfors regularity.

Lemma 4.6.18 *Let* $\Omega = \{(\Omega_m, E_m)\}_{m \geq 0}$ *be a proper system of horizontal networks and let* $N \in \mathbb{N}$. *If* $\underline{\mathcal{E}}_p(N_1, N_2, \Omega) = 0$, *then for any* $\eta > 0$ *and* $k_0 \in \mathbb{N}$, *there exist* $k_* \geq k_0$ *and* $\varphi : T^{(k_*)} \backslash \{\phi\} \to [0, 1]$ *such that, for any* $w \in T^{(k_*)}$,

$$\sum_{i=1}^m \varphi(w(i)) \geq 1 \tag{4.6.11}$$

for any $(w(1), \ldots, w(m)) \in C_{w,k_*}(N_1, N_2, N)$ *and*

$$\sum_{v \in S^{k_*}(w)} \varphi(v)^p < \eta. \tag{4.6.12}$$

Proof As $\underline{\mathcal{E}}_p(N_1, N_2, \Omega) = 0$ for any $\eta_0 > 0$ and $k_0 \in \mathbb{N}$, there exists $k_* \geq k_0$ such that $\mathcal{E}_{p,k_*,w}(N_1, N_2, \Omega) < \eta_0$ for any $w \in T$. Hence there exists $f_w : (T)_{|w|+k_*} \to [0, 1]$ such that $f_w(x) = 1$ for any $x \in \Omega^{k_*}(w, N_1)$, $f_w(x) = 0$ for any $x \in \Omega^{k_*,c}(w, N_2)$ and

$$\sum_{(x,y) \in E_{|w|+k_*}} |f_w(x) - f_w(y)|^p < \eta_0.$$

Let (N_0, L_0, L_1, L_2) be the indices of Ω. Set $n_0 = \min\{n | N \leq nL_1\}$ and $\overline{N} = (n_0 - 1)L_1 + L_2$. Note that $N \leq \overline{N}$ because $L_1 \leq L_2$. Define $E_m(U) = \cup_{u_1,u_2 \in U} E_m(u_1, u_2)$ for $U \subseteq (T)_m$. Define $\varphi_w : (T)_{|w|+k_*} \to [0, 1)$ by

$$\varphi_w(v) = \begin{cases} \displaystyle\sum_{(x,y) \in E_{|w|+k_*}(\Gamma_{\overline{N}}(v))} |f_w(x) - f_w(y)| & \text{if } v \in S^{k_*}(\Gamma_{N_2}(w)), \\ 0 & \text{otherwise.} \end{cases}$$

Let $(w(1), \ldots, w(m)) \in \mathcal{C}_{w,k_*}(N_1, N_2, N)$. By definition, there exist $w(0)$ and $w(m + 1) \in (T)_{|w|+k_*}$ such that $w(0) \in S^{k_*}(\Gamma_{N_1}(w))$, $w(m + 1) \notin S^{k_*}(\Gamma_{N_2}(w))$ and $(w(0), w(1), \ldots, w(m), w(m + 1))$ is a horizontal N-jpath. Choose $x_i \in \Omega_{|w|+k_*,w(i)}$ for $i = 0, \ldots, m + 1$. Note that $N \leq n_0 L_1$. By Lemma 4.6.14, for any $i = 1, \ldots, m$, there exist $(x_1^i, \ldots, x_{l_i}^i)$ and $(w^i(1), \ldots, w^i(l_i))$ such that $x_1^i = x_i$, $x_{l_i}^i = x_{i+1}$, $(x_j^i, x_{j+1}^i) \in E_{|w|+k_*}(w^i(j), w^i(j+1))$ and $w^i(j) \in \Gamma_{\overline{N}}(w(i))$ for any $j = 1, \ldots, l_i - 1$. For $i = 0$, again by Lemma 4.6.14, there exist $(x_1^0, \ldots, x_{l_0}^0)$ and $(w^0(1), \ldots, w^0(l_0))$ such that $x_1^0 = x_0$, $x_{l_0}^0 = x_1$, $(x_j^0, x_{j+1}^0) \in E_{|w|+k_*}(w^0(j), w^0(j+1))$ and $w^0(j) \in \Gamma_{\overline{N}}(w(1))$ for any $j = 1, \ldots, l_0 - 1$. Now, removing loops from $(x_0, x_2^0, \ldots, x_{l_0-1}^0, x_1, x_2^1, \ldots, x_{l_1-1}^1, x_2)$, we have a simple path $(z_1^1, z_2^1, \ldots, z_{m_1}^1)$ with $z_1^1 = x_0$ and $z_{m_1}^1 = x_2$. Since each (z_j^1, z_{j+1}^1) belongs to $E_{|w|+k_*}(\Gamma_{\overline{N}}(w(1)))$,

$$\varphi_w(w(1)) \geq \sum_{j=1}^{m_1-1} |f_w(z_j^1) - f_w(z_{j+1}^1)|$$

$$\geq \sum_{j=1}^{m_1-1} (f_w(z_j^1) - f_w(z_{j+1}^1)) = f_w(x_0) - f_w(x_2).$$

Applying the similar arguments to $(x_1^i, \ldots, x_{l_i}^i)$, we obtain

$$\varphi_w(w(i)) \geq f_w(x_i) - f_w(x_{i+1})$$

for any $i = 2, \ldots, m$. Hence

$$\sum_{i=1}^{m} \varphi_w(w(i)) \geq f(x_0) - f(x_{m+1}) \geq 1. \tag{4.6.13}$$

Next, we are going to estimate $\sum_{v \in (T)_{|w|+k_*}} \varphi_w(v)^p$. Since

$$\#(\{u \,|\, x \in \Omega_{|w|+k_*, u}\}) \leq L_*,$$

we see $\#(\{(u_1, u_2) \,|\, (x, y) \in E_m(u_1, u_2)\}) \leq (L_*)^2$ for any $(x, y) \in E_m$. Making use of this fact, we see that

$$\#(\{v \,|\, (x, y) \in E_{|w|+k_*}(\Gamma_{\overline{N}}(v))\}) \leq \sum_{(u_1, u_2) : (x, y) \in E_{|w|+k_*}(u_1, u_2)} \#(\Gamma_{\overline{N}}(u_1) \cap \Gamma_{\overline{N}}(u_2))$$

$$\leq \sum_{(u_1, u_2) : (x, y) \in E_{|w|+k_*}(u_1, u_2)} (L_*)^{\overline{N}} \leq (L_*)^{\overline{N}+2}.$$

Furthermore, $\#(\{(x, y) \,|\, (x, y) \in E_{|w|+k_*}(\Gamma_{\overline{N}}(v))\}) \leq (L_*)^{2\overline{N}} L_0$. Hence

$$\sum_{v \in (T)_{|w|+k_*}} \varphi_w(v)^p = \sum_{v \in (T)_{|w|+k_*}} \left(\sum_{(x,y) \in E_{|w|+k_*}(\Gamma_{\overline{N}}(v))} |f_w(x) - f_w(y)| \right)^p$$

$$\leq C_0 \sum_{v \in (T)_{|w|+k_*}} \sum_{(x,y) \in E_{|w|+k_*}(\Gamma_{\overline{N}}(v))} |f_w(x) - f_w(y)|^p$$

$$\leq C_0 \sum_{(x,y) \in E_{|w|+K_*}} \#(\{v \,|\, (x, y) \in E_{|w|+k_*}(\Gamma_{\overline{N}}(v))\}) |f_w(x) - f_w(y)|^p$$

$$\leq C_0 (L_*)^{\overline{N}+2} \eta_0,$$

where $C_0 = C_h((L_*)^{2\overline{N}} L_0, p)$. Define $\varphi : T^{(k_*)} \backslash \{\phi\} \to [0, 1]$ by

$$\varphi(v) = \max\{\varphi_w(v) \,|\, w \in (T)_{k_*(|v|-1)}\}.$$

By (4.6.13), we obtain (4.6.11). Since if $\varphi_w(v) > 0$, then $\pi^{k_*}(v) \in \Gamma_{N_2}(w)$, it follows that

$$\varphi(v) = \max\{\varphi_w(v) \,|\, w \in \Gamma_{N_2}(\pi^{k_*}(v))\}.$$

Therefore

$$\sum_{v \in S^{k*}(w)} \varphi(v)^p \leq \sum_{v \in S^{k*}(w)} \sum_{w' \in \Gamma_{N_2}(w)} \varphi_{w'}(v)^p$$

$$< C_0 \#(\Gamma_{N_2}(w))(L_*)^{\overline{N}+2} \eta_0 \leq C_0 (L_*)^{N_2+\overline{N}+2} \eta_0.$$

So, letting $\eta_0 = (C_0)^{-1}(L_*)^{-(N_2+\overline{N}+2)} \eta$, we verify (4.6.12).   □

Finally, we are going to complete the Proof of Theorem 4.6.9.

Proof of Theorem 4.6.9 Suppose $N_2 \geq N_1 + M_*$. By Lemma 4.6.16, it follows that $\overline{I}_{\mathcal{E}}(N_1, N_2, \Omega) \leq \dim_{AR}(X, d)$. To prove the opposite inequality, we assume that $\underline{I}_{\mathcal{E}}(0, N_2, \Omega.) < p$. Set $k_0 = \max\{m_0, k_{N_2}, k_{M_*}\}$ and $N = N_2$. Since $\mathcal{E}_p(0, N_2, \Omega) = 0$, Lemma 4.6.18 yields a function φ satisfying the assumptions of Theorem 4.5.1. Hence by Theorem 4.5.1, there exist a family of metrics $\{\rho_t\}_{t \in [0,1]}$ and a family of measures $\{\mu_t\}_{t \in [0,1]}$ such that $\rho_t \underset{QS}{\sim} d$ and μ_t is p-Ahlfors regular with respect to ρ_t for any $t \in [0, 1]$, and ρ_t and ρ_s are not bi-Lipschitz equivalent if $t \neq s$. This immediately shows that $\dim_{AR}(X, d) \leq p$. Hence we obtain $\dim_{AR}(X, d) \leq \underline{I}_{\mathcal{E}}(0, N_2, \Omega) \leq \underline{I}_{\mathcal{E}}(N_1, N_2, \Omega)$. Thus we have obtained

$$\dim_{AR}(X, d) \leq \underline{I}_{\mathcal{E}}(N_1, N_2, \Omega) \leq \overline{I}_{\mathcal{E}}(N_1, N_2, \Omega) \leq \dim_{AR}(X, d).$$

□

The proof of Corollary 4.6.10 is included in the above proof of Theorem 4.6.9.

Proof of Corollary 4.6.13 Applying Lemma 4.6.15 in the case where $\rho = d$, we obtain

$$\mathcal{E}_{p,k,w}(N_1, N_2, \Omega) \leq c \sum_{u \in S^k(\Gamma_{N_2}(w))} \left(\frac{d(u)}{d(w)}\right)^p \leq c' \#(S^k(\Gamma_{N_2}(w))) r^{pk}.$$

Set $\overline{N}_k(M) = \sup_{w \in T} \#(S^k(\Gamma_M(w)))$. Then

$$\mathcal{E}_{p,k}(N_1, N_2, \Omega) \leq c' \overline{N}_k(N_2) r^{pk}. \tag{4.6.14}$$

If $q - \epsilon > \liminf_{k \to \infty} -\frac{\log \overline{N}_k(N_2)}{k \log r}$, then there exists $\{k_j\}_{j \geq 1}$ such that

$$r^{\epsilon k_j} \geq \overline{N}_{k_j}(N_2) r^{q k_j}.$$

Hence by (4.6.14), $I_{\mathcal{E}}(N_1, N_2, \Omega) \leq q$. This implies

$$I_{\mathcal{E}}(N_1, N_2, \Omega) \leq \liminf_{k \to \infty} -\frac{\log \overline{N}_k(N_2)}{k \log r}.$$

By the definition of $\underline{N}_*$, we have the desired conclusion.

$\square$

4.7 Relation with p-Spectral Dimensions

By Theorem 4.6.9, we see that

$$\lim_{k \to \infty} \mathcal{E}_{p,k}(N_1, N_2, \Omega) = 0$$

if $p > \dim_{AR}(X, d)$ and

$$\liminf_{k \to \infty} \mathcal{E}_{p,k}(N_1, N_2, \Omega) > 0$$

if $p < \dim_{AR}(X, d)$. So, how about the rate of decay and/or divergence of $\mathcal{E}_{p,k}(N_1, N_2, \Omega)$ as $k \to \infty$? In this section, we define and investigate the rates and present another characterization of the Ahlfors regular conformal dimension in terms of them.

As in the previous sections, $(T, \mathcal{A}, \phi)$ is a locally finite tree with the root ϕ, $(X, \mathcal{O})$ is a compact metrizable topological space with no isolated point, $K : T \to \mathcal{C}(X, \mathcal{O})$ is a minimal partition. We also assume that $\sup_{w \in T} \#(S(w)) < +\infty$. Furthermore we fix $d \in \mathcal{D}_{A,\epsilon}(X, \mathcal{O})$ satisfying (BF1) and (BF2) in Sect. 4.3.

In this framework, the rates are defined as follows.

Definition 4.7.1 Let $\Omega = \{(\Omega_m, E_m)\}_{m \geq 0}$ be a proper system of horizontal networks. For $p > 0$, define

$$\overline{R}_p(N_1, N_2, \Omega) = \limsup_{n \to \infty} \mathcal{E}_{p,n}(N_1, N_2, \Omega)^{\frac{1}{n}}$$

$$\underline{R}_p(N_1, N_2, \Omega) = \liminf_{n \to \infty} \mathcal{E}_{p,n}(N_1, N_2, \Omega)^{\frac{1}{n}}.$$

First of all, $\overline{R}_p$ and $\underline{R}_p$ are finite as one can see in the next proposition.

Proposition 4.7.2 Let $\Omega = \{(\Omega_m, E_m)\}_{m \geq 0}$ be a proper system of horizontal networks. Let $N_2 \geq N_1 + M_*$. Then for any $p > 0$,

$$\underline{R}_p(N_1, N_2, \Omega) \leq \overline{R}_p(N_1, N_2, \Omega) \leq r^p \overline{N}_*.$$

Proof By (4.6.14),

$$\mathcal{E}_{p,k}(N_1, N_2, \Omega) \leq c' \sup_{w \in T} \#(S^k(\Gamma_{N_2}(w))) r^{pk}.$$

This immediately implies the desired inequality.

$$\square$$

Ideally, we expect that $\overline{R}_p(N_1, N_2, \Omega) = \underline{R}_p(N_1, N_2, \Omega)$ and that

$$\frac{1}{(R_p)^m} \mathcal{E}_p(f | \Omega, E_m) \to \mathcal{E}_p(f) \quad \text{as } m \to \infty,$$

where $R_p = \overline{R}_p(N_1, N_2, \Omega)$, for f belonging to some reasonably large class of functions, although the expression $\mathcal{E}_p(f)$ makes little sense for now. As a matter of fact, for a class of (random) self-similar sets including the Sierpinski gasket and the (generalized) Sierpinski carpets, it is known that

$$\overline{R}_2(N_1, N_2, \Omega) = \underline{R}_2(N_1, N_2, \Omega). \tag{4.7.1}$$

Moreover, the rate R_2 is called the resistance scaling ratio and $\mathcal{E}_2(f)$ has known to induce the "Brownian motion" $(\{X_t\}_{t>0}, \{P_x\}_{x \in X})$ and the "Laplacian" Δ through the formula

$$\mathcal{E}_2(f) = \int_X f \Delta f d\mu$$

$$E_x(f(X_t)) = (e^{-t\Delta} f)(x),$$

where $E_x(\cdot)$ is the expectation with respect to $P_x(\cdot)$. See [2, 3, 5, 24] and [19] for details

Now we start to study the relation between R_p's for different values of p.

Lemma 4.7.3 *Let* $\Omega = \{(\Omega_m, E_m)\}_{m \geq 0}$ *be a proper system of horizontal networks. If* $0 < p < q$, *then there exists* $c > 0$ *such that*

$$\mathcal{E}_{p,k}(N_1, N_2, \Omega)^{\frac{1}{p}} \leq c \mathcal{E}_{q,k}(N_1, N_2, \Omega)^{\frac{1}{q}} \sup_{w \in T} \#(S^k(\Gamma_{N_2}(w)))^{\frac{1}{p} - \frac{1}{q}}$$

for any $k \geq 1$.

Proof Let $(Y, \mathcal{M}, \mu)$ is a measurable space. Assume $\mu(Y) < \infty$. Then by the Hölder inequality,

$$\int_Y |u|^p d\mu \leq \left(\int_Y |u|^q d\mu \right)^{\frac{p}{q}} \mu(Y)^{\frac{q-p}{q}}.$$

Applying this to $\mathcal{E}_p(f|\Omega_{|w|+k}, E_{|w|+k})$, we obtain

$$\mathcal{E}_p(f|\Omega_{|w|+k}, E_{|w|+k})^{\frac{1}{p}}$$

$$\leq \mathcal{E}_q(f|\Omega_{|w|+k}, E_{|w|+k})^{\frac{1}{q}} \left(\#(\{(x, y)|(x, y) \in E_{|w|+k}, x \in \Omega^k(w, N_2)\}) \right)^{\frac{1}{p}-\frac{1}{q}}$$

$$\leq \mathcal{E}_q(f|\Omega_{|w|+k}, E_{|w|+k})^{\frac{1}{q}} \left(\sum_{u \in S^k(\Gamma_{N_2}(w))} \sum_{v \in \Gamma_N(u)} \#(E_m(u, v)) \right)^{\frac{1}{p}-\frac{1}{q}}$$

$$\leq \mathcal{E}_q(f|\Omega_{|w|+k}, E_{|w|+k})^{\frac{1}{q}} \left(L_0(L_*)^N \#(S^k(\Gamma_{N_2}(w))) \right)^{\frac{1}{p}-\frac{1}{q}}$$

for any $f \in \mathcal{F}_F(\Omega_{|w|+k}, \Omega^k(w, N_1), \Omega^{k,c}(w, N_2))$. Set $c = \left(L_0(L_*)^N \right)^{\frac{1}{p}-\frac{1}{q}}$. Then the above inequality implies that

$$\mathcal{E}_{p,k,w}(N_1, N_2, \Omega)^{\frac{1}{p}} \leq c \mathcal{E}_{q,k,w}(N_1, N_2, \Omega)^{\frac{1}{q}} \#(S^k(\Gamma_{N_2}(w)))^{\frac{1}{p}-\frac{1}{q}}.$$

This shows the desired inequality. $\qquad\square$

Lemma 4.7.3 immediately implies the following fact.

Lemma 4.7.4 *Let* $\Omega = \{(\Omega_m, E_m)\}_{m \geq 0}$ *be a proper system of horizontal networks. If* $0 < p < q$, *then*

$$\overline{R}_p(N_1, N_2, \Omega)^{\frac{1}{p}} \leq \overline{R}_q(N_1, N_2, \Omega)^{\frac{1}{q}} (\overline{N}_*)^{\frac{1}{p}-\frac{1}{q}} \tag{4.7.2}$$

$$\underline{R}_p(N_1, N_2, \Omega)^{\frac{1}{p}} \leq \underline{R}_q(N_1, N_2, \Omega)^{\frac{1}{q}} (\overline{N}_*)^{\frac{1}{p}-\frac{1}{q}}. \tag{4.7.3}$$

Using this lemma, we can show the continuity and the monotonicity of $\underline{R}_p$ and $\overline{R}_p$.

Proposition 4.7.5 *Let* $\Omega = \{(\Omega_m, E_m)\}_{m \geq 0}$ *be a proper system of horizontal networks. Let* $N_2 \geq N_1 + M_*$.

(1) $\overline{R}_p(N_1, N_2, \Omega)$ *and* $\underline{R}_p(N_1, N_2, \Omega)$ *are continuous and monotonically non-increasing as a function of* p.

(2) $\overline{R}_p(N_1, N_2, \Omega) < 1$ *for any* $p > \dim_{AR}(X, d)$.

Proof (1) Since $|f(x) - f(y)| \leq 1$ if $0 \leq f(x) \leq 1$ for any $x \in \Omega_m$, we see that

$$\mathcal{E}_{p,n,w}(N_1, N_2, \Omega) \geq \mathcal{E}_{q,n,w}(N_1, N_2, \Omega)$$

whenever $p < q$. Hence $\underline{R}_p$ and $\overline{R}_p$ are monotonically non-increasing. Set $R_p = \overline{R}_p(N_1, N_2, \Omega)$. By (4.7.2), if $p < q$, then

$$R_q \le R_p \le (R_q)^{\frac{p}{q}} (\overline{N}_*)^{1-\frac{p}{q}}.$$

This shows that $\lim_{p\uparrow q} R_p = R_q$. Exchanging p and q, we obtain

$$(R_q)^{\frac{p}{q}} (\overline{N}_*)^{1-\frac{p}{q}} \le R_p \le R_q$$

if $q < p$. This implies $\lim_{p\downarrow q} R_p = R_q$. Thus R_p is continuous. The same discussion works for $\underline{R}_p(N_1, N_2, \Omega)$ as well.

(2) This follows from (4.6.10).   □

Consequently, we obtain another characterization of the AR conformal dimension.

Theorem 4.7.6 *Let* $\Omega = \{(\Omega_m, E_m)\}_{m\ge 0}$ *be a proper system of horizontal networks. Assume that* $N_2 \ge N_1 + M_*$. *Then*

$$\dim_{AR}(X, d) = \inf\{p|\overline{R}_p(N_1, N_2, \Omega) < 1\} = \max\{p|\overline{R}_p(N_1, N_2, \Omega) = 1\}$$
$$= \inf\{p|\underline{R}_p(N_1, N_2, \Omega) < 1\} = \max\{p|\underline{R}_p(N_1, N_2, \Omega) = 1\}.$$
$$(4.7.4)$$

In particular, if $p_* = \dim_{AR}(X, d)$, *then*

$$\lim_{n\to\infty} \mathcal{E}_{p_*,n}(N_1, N_2, \Omega)^{\frac{1}{n}} = 1. \qquad (4.7.5)$$

Proof Write $\underline{R}_p = \underline{R}_p(N_1, N_2, \Omega)$ and $p_* = \dim_{AR}(X, d)$. Since $\underline{R}_p$ is continuous, $\lim_{p\downarrow p_*} \underline{R}_p \le 1$. If this limit is less than 1, the continuity of $\underline{R}_p$ implies that $\underline{R}_{p_*-\epsilon} < 1$ for sufficiently small $\epsilon > 0$. Then $\underline{\mathcal{E}}_{p_*-\epsilon}(N_1, N_2, \Omega) = 0$. Hence $p_* - \epsilon \ge p_*$. This contradiction shows that $\lim_{p\downarrow p_*} \underline{R}_p = 1$. Consequently, $\underline{R}_{p_*} = 1$. Since $\underline{R}_p < 1$ if $p > p_*$, we may verify (4.7.4) for $\underline{R}_p(N_1, N_2, \Omega)$. The same discussion works for $\overline{R}_p(N_1, N_2, \Omega)$ as well. Consequently, $\overline{R}_{p_*}(N_1, N_2, \Omega) = \underline{R}_{p_*}(N_1, N_2, \Omega) = 1$. Hence we have (4.7.5)   □

Next we define the (upper and lower) p-spectral dimension $\overline{d}_p^S(N_1, N_2, \Omega)$ and $\underline{d}_p^S(N_1, N_2, \Omega)$.

Definition 4.7.7 For $p > 0$, define $\overline{d}_p^S(N_1, N_2, \Omega)$ and $\underline{d}_p^S(N_1, N_2, \Omega)$ by

$$\overline{d}_p^S(N_1, N_2, \Omega) = \frac{p \log \overline{N}_*}{\log \overline{N}_* - \log \overline{R}_p(N_1, N_2, \Omega)}$$

$$\underline{d}_p^S(N_1, N_2, \Omega) = \frac{p \log \overline{N}_*}{\log \overline{N}_* - \log \underline{R}_p(N_1, N_2, \Omega)}.$$

The quantities $\overline{d}_p^S(N_1, N_2, \Omega)$ and $\underline{d}_p^S(N_1, N_2, \Omega)$ are called the upper p-spectral dimension and the lower p-spectral dimension respectively.

Note that $\overline{d}_p^S(N_1, N_2, \Omega)$ and $\underline{d}_p^S(N_1, N_2, \Omega)$ coincide with the unique numbers $\overline{d}, \underline{d} \in \mathbb{R}$ satisfying

$$\overline{N}_* \left(\frac{\overline{R}_p(N_1, N_2, \Omega)}{\overline{N}_*} \right)^{\overline{d}/p} = 1 \quad \text{and} \quad \overline{N}_* \left(\frac{\underline{R}_p(N_1, N_2, \Omega)}{\overline{N}_*} \right)^{\underline{d}/p} = 1 \qquad (4.7.6)$$

respectively.

For the Sierpinski gasket and the generalized Sierpinski carpets, the equality (4.7.1) implies $\overline{d}_2^S(N_1, N_1, \Omega) = \underline{d}_2^S(N_1, N_2, \Omega)$, which is called the spectral dimension and written as d^S. The spectral dimension has been known to represent asymptotic behaviors of the Brownian motion and the Laplacian. See [4, 5] and [23] for example. More precisely, if $p(t, x, y)$ is the transition density of the Brownian motion, then

$$c_1 t^{-d^S/2} \leq p(t, x, x) \leq c_2 t^{-d^S/2}$$

for any $t \in (0, 1]$ and $x \in X$. Furthermore, let $N(\cdot)$ is the eigenvalue counting function of Δ, i.e.

$$N(\lambda) = \text{the number of eigenvalues } \leq \lambda \text{ taking the multiplicity into account.}$$

Then

$$c_1 \lambda^{d^S/2} \leq N(\lambda) \leq c_2 \lambda^{d^S/2}$$

for any $\lambda \geq 1$.

Immediately by the above definition, we obtain the following lemma.

Lemma 4.7.8

(a) $\overline{R}_p(N_1, N_2, \Omega) < 1$ if and only if $\overline{d}_p^S(N_1, N_2, \Omega) < p$,

(b) $\overline{R}_p(N_1, N_2, \Omega) = 1$ if and only if $\overline{d}_p^S(N_1, N_2, \Omega) = p$,

(c) $\overline{R}_p(N_1, N_2, \Omega) > 1$ if and only if $\overline{d}_p^S(N_1, N_2, \Omega) > p$.

The same relations hold between $\underline{R}_p(N_1, N_2, \Omega)$ *and* $\underline{d}_p^S(N_1, N_2, \Omega)$ *as well.*

Finally we present the relation between p-spectral dimension and the Ahlfors regular conformal dimension.

Theorem 4.7.9 *Let* $\Omega = \{(\Omega_m, E_m)\}_{m\geq 0}$ *be a proper system of horizontal networks. Assume that* $N_2 \geq N_1 + M_*$. *Then for* $p > 0$, *either of the following two cases occurs:*

(1) $\overline{R}_p(N_1, N_2, \Omega) < 1$ *and* $\underline{R}_p(N_1, N_2, \Omega) < 1$. *In this case*

$$\dim_{AR}(X, d) \leq \underline{d}_p^S(N_1, N_2, \Omega) \leq \overline{d}_p^S(N_1, N_2, \Omega) < p.$$

(2) $\overline{R}_p(N_1, N_2, \Omega) \geq 1$ *and* $\underline{R}_p(N_1, N_2, \Omega) \geq 1$. *In this case*

$$\dim_{AR}(X, d) \geq \overline{d}_p^S(N_1, N_2, \Omega) \geq \underline{d}_p^S(N_1, N_2, \Omega) \geq p.$$

Note that the above two cases are not compatible.

For the case of the Sierpinski gasket and the (generalized) Sierpinski carpets, the above theorem shows that either

$$\dim_{AR}(X, d_*) \leq d^S < 2$$

or

$$\dim_{AR}(X, d_*) \geq d^S \geq 2,$$

where d_* is the restriction of the Euclidean metric. For the standard planar Sierpinski carpet in Example 4.6.7, it has been shown in [6] that

$$d^S \leq 1.805$$

by rigorous numerical estimate. This gives an upper estimate of the Ahlfors regular conformal dimension of the Sierpinski carpet.

Proof of Theorem 4.7.9 Write $\underline{R}_p = \underline{R}_p(N_1, N_2, \Omega)$, $\overline{R}_p = \overline{R}(N_1, N_2, \Omega)$, $\underline{d}_p = \underline{d}_p^S(N_1, N_2, \Omega)$ and $\overline{d}_p = \overline{d}_p^S(N_1, N_2, \Omega)$. First we examine 6 cases.

Case 1 $\underline{R}_p < 1$: By Lemma 4.7.8, we have $\underline{d}_p < p$. Let $q \in (\underline{d}_p, p)$. By (4.7.6), there exists $\epsilon > 0$ such that

$$(1 + \epsilon)\overline{N}_* \left(\frac{\underline{R}_p}{\overline{N}_*}\right)^{q/p} < 1.$$

Choose $\{n_j\}_{j \geq 1}$ so that $\underline{R}_p = \lim_{j \to \infty} \mathcal{E}_{p,n_j}(N_1, N_2, \Omega)^{\frac{1}{n_j}}$. Then for sufficiently large j, we have

$$\mathcal{E}_{p,n_j}(N_1, N_2, \Omega) \leq ((1 + \epsilon)\underline{R}_p)^{n_j} \quad \text{and} \quad \sup_{w \in T} \#(S^{n_j}(\Gamma_{N_2}(w)))$$

$$\leq ((1 + \epsilon)\overline{N}_*)^{n_j}.$$

Hence by Lemma 4.7.3, as $j \to 0$,

$$\mathcal{E}_{q,n_j}(N_1, N_2, \Omega) \leq c\left((1 + \epsilon)\overline{N}_*\left(\frac{\underline{R}_p}{\overline{N}_*}\right)^{\frac{q}{p}}\right)^{n_j} \to 0.$$

Therefore $\underline{I}_{\mathcal{E}}(0, M_1, \Omega) \leq q$. Using Theorem 4.6.9, we have $\dim_{AR}(X, d) \leq \underline{d}_p$. Consequently, we see

$$\dim_{AR}(X, d) \leq \underline{d}_p < p.$$

Case 2 $\overline{R}_p < 1$: Similar arguments as in Case 1 shows

$$\dim_{AR}(X, d) \leq \overline{d}_p < p.$$

Moreover, since $\underline{R}_p \leq \overline{R}_p$, Case 2 is included by Case 1 and so

$$\dim_{AR}(X, d) \leq \underline{d}_p \leq \overline{d}_p < p.$$

Case 3 $\overline{R}_p > 1$: By Lemma 4.7.8, we have $\overline{d}_p > p$. Let $q \in (p, \overline{d}_p)$. By (4.7.6),

$$\frac{\overline{N}_*}{1 - \epsilon}\left((1 - \epsilon)^2 \frac{\overline{R}_p}{\overline{N}_*}\right)^{\frac{q}{p}} > 1$$

for sufficiently small $\epsilon > 0$. Choose $\{n_j\}_{j \geq 1}$ so that $\mathcal{E}_{p,n_j}(N_1, N_2, \Omega)^{\frac{1}{n_j}} \to \overline{R}_p$ as $j \to \infty$. Then for sufficiently large j, we have

$$(1 - \epsilon)\overline{R}_p \leq \mathcal{E}_{p,n_j}(N_1, N_2, \Omega)^{\frac{1}{n_j}} \quad \text{and} \quad \sup_{w \in T} \#(S^{n_j}(\Gamma_{N_2}(w)))$$

$$\leq \left(\frac{\overline{N}_*}{1 - \epsilon}\right)^{n_j}.$$

Using Lemma 4.7.3, we have

$$((1 - \epsilon)\overline{R}_p)^{n_j \frac{q}{p}} \leq c\mathcal{E}_{q,n_j}(N_1, N_2, \Omega)\left(\frac{\overline{N}_*}{1 - \epsilon}\right)^{n_j \frac{q-p}{p}}$$

for sufficiently large j. This implies

$$1 < \left(\frac{\overline{N}_*}{1 - \epsilon}\left((1 - \epsilon)^2 \frac{\overline{R}_p}{\overline{N}_*}\right)^{\frac{q}{p}}\right)^{n_j} \leq c\mathcal{E}_{q,n_j}(N_1, N_2, \Omega).$$

Thus we have $\overline{R}_q > 1$. Hence $q \leq \dim_{AR}(X, d)$ by Theorem 4.7.6. Consequently

$$\dim_{AR}(X, d) \geq \overline{d}_p > p.$$

Case 4 $\underline{R}_p > 1$: Similar arguments as in Case 3 shows

$$\dim_{AR}(X, d) \geq \underline{d}_p > p.$$

Moreover, since $\overline{R}_p \geq \underline{R}_p$, this case is included by Case 3 and hence

$$\dim_{AR}(X, d) \geq \overline{d}_p \geq \underline{d}_p > p.$$

Case 5 $\overline{R}_p = 1$: By Theorem 4.7.6,

$$\dim_{AR}(X, d) \geq \overline{d}_p = p.$$

Case 6 $\underline{R}_p = 1$: By Theorem 4.7.6,

$$\dim_{AR}(X, d) \geq \underline{d}_p = p.$$

Now that we have investigated the above six cases, the desired statement is straightforward. In fact, if $\underline{R}_p < 1$, then $\dim_{AR}(X, d) < p$. This leaves us only one choice, Case 2, i.e. $\overline{R}_p < 1$. Similarly, we see that $\overline{R}_p \geq 1$ if and only if $\underline{R}_p \geq 1$. □

4.8 Combinatorial Modulus of Curves

Originally in [13], the characterization of the Ahlfors regular conformal dimension has been given in terms of the critical exponent of p-combinatorial modulus of curve families. In this section, we are going to show a direct correspondence between p-energies and p-combinatorial moduli and reproduce Piaggio's result in [13] within our framework.

As in the previous sections, $(T, \mathcal{A}, \phi)$ is a locally finite tree with the root ϕ, $(X, \mathcal{O})$ is a compact metrizable topological space with no isolated point, $K : T \to \mathcal{C}(X, \mathcal{O})$ is a minimal partition. We also assume that $\sup_{w \in T} \#(S(w)) < +\infty$. Furthermore, we fix $d \in \mathcal{D}_{A,\epsilon}(X, \mathcal{O})$ satisfying the basic framework, i.e. (BF1) and (BF2) in Sect. 4.3.

Definition 4.8.1 Let (V, E) be a non-directed graph. Set

$$\mathcal{P}(V, E) = \{(x(1), \ldots, x(n)) | x(i) \in V \text{ for any } i = 1, \ldots, n \text{ and}$$

$$(x(i), x(i+1)) \in E \text{ for any } i = 1, \ldots, n-1\}.$$

For $U_1, U_2 \subseteq V$ with $U_1 \cap U_2 = \emptyset$, set

$$\mathcal{C}(V, E, U_1, U_2) = \{(x(1), \ldots, x(m)) | \text{ there exist } x(0) \in U_1 \text{ and}$$

$$x(m+1) \in U_2 \text{ such that } (x(0), \ldots, x(m+1)) \in \mathcal{P}(V, E)\}.$$

Define

$$\mathcal{F}_M(V, E, U_1, U_2) = \{f | f : V \to [0, \infty), \sum_{i=1}^{m} f(x(i)) \geq 1$$

$$\text{for any } (x(1), \ldots, x(m)) \in \mathcal{C}(V, E, U_1, U_2)\}$$

and for $p > 0$,

$$\mathrm{Mod}_p(V, E, U_1, U_2) = \inf\{\sum_{x \in V} |f(x)|^p | f \in \mathcal{F}_M(V, E, U_1, U_2)\},$$

which is called the *p-modulus of curves connecting U_1 and U_2.*

Definition 4.8.2 Let (V, E) be a non-directed graph. Assume that $U_1, U_2 \subseteq V$ and $U_1 \cap U_2 = \emptyset$.

(1) For $f \in \mathcal{F}_M(V, E, U_1, U_2)$, define $F(f) : V \to [0, \infty)$ as

$$F(f)(x) = \min \left\{ \sum_{i=1}^{k} f(x(i)) \middle| k \geq 1, (x(0), \ldots, x(k)) \right.$$

$$\left. \in \mathcal{P}(V, E), x(0) \in U_2, x(k) = x \right\}.$$

for $x \notin U_2$ and $F(f)(x) = 0$ for $x \in U_2$.

(2) For $g \in \mathcal{F}_F(V, U_1, U_2)$, define

$$G(g)(x) = \sum_{(x,y) \in E} |g(x) - g(y)|.$$

Lemma 4.8.3 *Let (V, E) be a non-directed graph. Assume that $U_1, U_2 \subseteq V$ and $U_1 \cap U_2 = \emptyset$. Define $L(V, E) = \max_{x \in V} \#(\{y | (x, y) \in E\})$.*

(1) *For any $f \in \mathcal{F}_M(V, E, U_1, U_2)$, $F(f) \in \mathcal{F}_F(V, U_1, U_2)$ and*

$$\mathcal{E}_p(F(f)|V, E) \leq L(V, E) \sum_{x \in V} f(x)^p. \tag{4.8.1}$$

(2) *For any $g \in \mathcal{F}_F(V, U_1, U_2)$, $G(g) \in \mathcal{F}_M(V, E, U_1, U_2)$ and*

$$\sum_{x \in V} G(g)(x)^p \leq 2C_h(p, L(E, V))\mathcal{E}_p(g|V, E). \tag{4.8.2}$$

Proof (1) The claim that $F(f) \in \mathcal{F}_F(V, U_1, U_2)$ is immediate by the definition. If $(x, y) \in E$, then

$$F(f)(x) + f(y) \geq F(f)(y) \quad \text{and} \quad F(f)(y) + f(x) \geq F(f)(x).$$

Therefore,

$$|F(f)(x) - F(f)(y)| \leq \max\{f(x), f(y)\}.$$

Thus

$$\mathcal{E}_p(F(f)|V, E) = \frac{1}{2} \sum_{(x,y) \in E} |F(f)(x) - F(f)(y)|^p$$

$$\leq \frac{1}{2} \sum_{(x,y) \in E} (f(x)^p + f(y)^p) \leq L(V, E) \sum_{x \in V} f(x)^p.$$

(2) Let $(x(1), \ldots, x(m)) \in \mathcal{C}(V, E, U_1, U_2)$. Then $(x(0), x(1)) \in E$ for some $x(0) \in U_1$ and $(x(m), x(m+1)) \in U_2$ for some $x(m+1) \in V_2$. Without loss of generality, we may assume that $(x(0), x(1), \ldots, x(m+1))$ does not contain any loop. Then

$$G(w(1)) \geq |g(w(0)) - g(w(1))| + |g(w(1)) - g(w(2))|$$

and

$$G(w(i)) \geq |g(w(i)) - g(w(i+1))|$$

for any $i = 2, \ldots, m$. Hence

$$\sum_{i=1}^{m} G(w(i)) \geq \sum_{j=0}^{m} |g(w(j)) - g(w(j+1))| \geq g(w(0)) - g(w(m+1)) \geq 1.$$

Thus $G(g) \in \mathcal{F}_M(V, E, U_1, U_2)$. Moreover, by Lemma 4.6.17

$$\sum_{x \in V} G(g)(x)^p \leq C_h(p, L(E, V)) \sum_{x \in V} \sum_{y:(x,y) \in E} |g(x) - g(y)|^p$$

$$\leq 2C_h(p, L(E, V))\mathcal{E}_p(g|V, E).$$

$\square$

Taking infimums in (4.8.1) and (4.8.2), we obtain the following proposition giving a direct connection between p-energy and p-modulus.

Proposition 4.8.4 *Let (V, E) be a non-directed graph. Assume that $U_1, U_2 \subseteq V$ and $U_1 \cap U_2 = \emptyset$. Then*

$$\mathcal{E}_p(V, E, U_1, U_2) \leq L(V, E)\mathrm{Mod}_p(V, E, U_1, U_2)$$

and

$$\mathrm{Mod}_p(V, E, U_1, U_2) \leq 2C_h(p, L(V, E))\mathcal{E}_p(V, E, U_1, U_2).$$

Next we give the definition of the critical index of p-moduli.

Definition 4.8.5 Let $\Omega = \{(\Omega_m, E_m)\}_{m \geq 0}$ be a proper system of horizontal networks. Define

$$\mathcal{M}_{p,k,w}(N_1, N_2, \Omega) = \mathrm{Mod}_p(\Omega_{|w|+k}, E_{|w|+k}, \Omega^k(w, N_1), \Omega^{k,c}(w, N_2)),$$

$$\mathcal{M}_{p,k}(N_1, N_2, \Omega) = \sup_{w \in T} \mathcal{M}_{p,k,w}(N_1, N_2, \Omega),$$

$$\overline{\mathcal{M}}_p(N_1, N_2, \Omega) = \limsup_{k \to \infty} \mathcal{M}_{p,k}(N_1, N_2, \Omega),$$

$$\underline{\mathcal{M}}_p(N_1, N_2, \Omega) = \liminf_{k \to \infty} \mathcal{M}_{p,k}(N_1, N_2, \Omega),$$

$$\overline{I}_\mathcal{M}(N_1, N_2, \Omega) = \inf\{p | \overline{\mathcal{M}}_p(N_1, N_2, \Omega) = 0\},$$

$$\underline{I}_\mathcal{M}(N_1, N_2, \Omega) = \inf\{p | \underline{\mathcal{M}}_p(N_1, N_2, \Omega) = 0\}.$$

Due to Proposition 4.8.4, $\mathcal{E}_{p,n}(N_1, N_2, \Omega)$ and $\mathcal{M}_{p,n}(N_1, N_2, \Omega)$ can be compared in the following way.

Lemma 4.8.6 *Let* $\Omega = \{(\Omega_m, E_m)\}_{m \geq 0}$ *be a proper system of horizontal networks with indices* (N, L_0, L_1, L_2). *Then*

$$\mathcal{E}_{p,n}(N_1, N_2, \Omega) \leq L_0(L_*)^{N+1}\mathcal{M}_{p,n}(N_1, N_2, \Omega)$$

and

$$\mathcal{M}_{p,n}(N_1, N_2, \Omega) \leq 2C_h(p, L_0(L_*)^{N+1})\mathcal{E}_{p,n}(N_1, N_2, \Omega).$$

Proof It is enough to show that $L(\Omega_m, E_m) \leq L_0(L_*)^{N+1}$. If $x \in V_m$, then by (N4),

$$\{y|y \in \Omega_m, (x, y) \in E_m\} \subseteq \bigcup_{w:x\in K_w} \bigcup_{v\in\Gamma_N(w)} \bigcup_{(x,y)\in E_m(w,v)} \{y\}.$$

Using (N3), we see that

$$\#(\{y|y \in \Omega_m, (x, y) \in E_m\}) \leq \#(\{w|x \in K_w\})\#(\Gamma_N(w)))L_0$$

$$\leq L_*(L_*)^N L_0.$$

If $x = w \in A_m = (T)_m \cap \Omega_m$, similar arguments show that $\#(y|y \in \Omega_m, (x, y) \in E_m\}) \leq (L_*)^N L_0$. Thus we have $L(\Omega, E_m) \leq (L_*)^{N+1}L_0$.   □

The above lemma combined with Theorem 4.6.9 yields the following characterization of the Ahlfors regular conformal dimension by the critical exponents of discrete moduli.

Theorem 4.8.7 *Let* $\Omega = \{(\Omega_m, E_m)\}_{m \geq 0}$ *be a proper system of horizontal networks. If* $N_2 \geq N_1 + M_*$, *then*

$$\overline{I}_{\mathcal{M}}(N_1, N_2, \Omega) = \underline{I}_{\mathcal{M}}(N_1, N_2, \Omega) = \dim_{AR}(X, d).$$

4.9 Positivity at the Critical Value

One of the advantages of the use of discrete moduli is to show the positivity of $\underline{\mathcal{M}}_p(N_1, N_2, \Omega)$ and $\underline{\mathcal{E}}_p(N_1, N_2, \Omega)$ at the critical value $p_* = \dim_{AR}(X, d)$.

As in the previous sections, $(T, \mathcal{A}, \phi)$ is a locally finite tree with the root ϕ, $(X, \mathcal{O})$ is a compact metrizable topological space with no isolated point, $K : T \to \mathcal{C}(X, \mathcal{O})$ is a minimal partition. We also assume that $\sup_{w\in T} \#(S(w)) < +\infty$. Furthermore, we fix $d \in \mathcal{D}_{A,\epsilon}(X, \mathcal{O})$ satisfying the basic framework, i.e. (BF1) and (BF2) in Sect. 4.3.

Theorem 4.9.1 *Let* $\Omega = \{(\Omega_m, E_m)\}_{m \geq 0}$ *be a proper system of horizontal networks. Suppose* $N_2 \geq N_1 + M_*$. *Let* $p_* = \dim_{AR}(X, d)$. *If* $p_* > 0$, *then*

$$\underline{\mathcal{M}}_{p_*}(N_1, N_2, \Omega) > 0 \quad and \quad \underline{\mathcal{E}}_{p_*}(N_1, N_2, \Omega) > 0.$$

First step of a proof is to modify the original proper system of horizontal networks.

Notation

$$\mathcal{Q}_{w,k}(M_1, M_2, \Omega) = \mathcal{F}_M(\Omega_{|w|+k}, E_{|w|+k},, \Omega^k(w, M_1), \Omega^{k,c}(w, M_2)) \qquad (4.9.1)$$

and

$$\widetilde{\mathcal{C}}_{w,k}(M_1, M_2, \Omega) = \mathcal{C}(\Omega_{|w|+k}, E_{|w|+k}, \Omega^k(w, M_1), \Omega^{k,c}(w, M_2)) \qquad (4.9.2)$$

for $M_1, M_2 \geq 1$. Note that $\widetilde{\mathcal{C}}_{w,k}(M_1, M_2, \Omega_*^{(M)}) = \mathcal{C}_{w,k}(M_1, M_2, M)$, where $\Omega_*^{(M)}$ is defined in Example 4.6.6 as $\Omega_*^{(M)} = \{((T)_m, J_{M,m}^h)\}_{m \geq 0}$.

Lemma 4.9.2 *Let* $\Omega = \{(\Omega_m, E_m)\}_{m \geq 0}$ *be a proper system of horizontal networks with indices* (N, L_0, L_1, L_2). *Define* $\mathcal{S}_m = \{w | w \in (T)_m, \Gamma_1(w) = \{w\}\}$.

(1) *If* $w \in (T)_m \backslash \mathcal{S}_m$, *then*

$$\#(\Omega_{m,w}) \leq L_0(L_*)^{N+1}.$$

(2) *For* $m \geq 0$ *and* $M \geq 1$, *define*

$$J_{M,m}^h[\Omega] = \bigcup_{w,v \in (T)_m, v \in \Gamma_M(w)} \Omega_{m,w} \times \Omega_{m,v}.$$

Set $\overline{\Omega}^M = \{(\Omega_m, J_{M,m}^h[\Omega])\}_{m \geq 0}$. *Then* $\overline{\Omega}^M$ *is a proper system of horizontal networks with indices* $(M, L_0^2(L_*)^{2N+4}, M, M)$. *Moreover, there exists* $c > 0$ *such that*

$$\mathcal{E}_p(f | \Omega_m, J_{M,m}^h[\Omega]) \leq c \mathcal{E}_p(f | \Omega_m, E_m) \qquad (4.9.3)$$

for any $m \geq 0$ *and* $f : \Omega_m \to \mathbb{R}$.

(3)

$$\mathcal{M}_{p,k,w}(N_1, N_2, \overline{\Omega}^M)$$

$$= \inf \left\{ \sum_{x \in \Omega_{|w|+k}(S^k(\Gamma_{N_2}(w)) \backslash \mathcal{S}_{|w|+k})} |f(x)|^p \, \middle| \, f \in \mathcal{Q}_{w,k}(N_1, N_2, \overline{\Omega}^M) \right\}.$$

Proof (1) Let $v \in \Gamma_1(w) \backslash \{w\}$. Since $(w, v) \in J^h_{L_1}$, (N5) implies that for any $x \in \Omega_{m,w}$, there exists $y \in \Omega_m$ such that $(x, y) \in E_m$. By (N4), there exists $u \in \Gamma_{N+1}(w)$ such that $y \in \Omega_{m,u}$. Hence

$$\{(x, y) | (x, y) \in E_m, x \in \Omega_{m,w}\} \subseteq \bigcup_{u \in \Gamma_{N+1}(w)} E_m(w, u).$$

and the projection from $\{(x, y) | (x, y) \in E_m, x \in \Omega_{m,w}\}$ to $\Omega_{m,w}$ is surjective. Therefore by (N3),

$$\#(\Omega_{m,w}) \leq \#(\{(x, y) | (x, y) \in E_m, x \in \Omega_{m,w}\})$$

$$\leq \sum_{u \in \Gamma_{N+1}(w)} \#(E_m(w, v)) \leq (L_*)^{N+1} L_0.$$

(2) Suppose that $(x, y) \in J^h_{M,m}[\Omega]$, $x \in \Omega_{m,w}$ and $y \in \Omega_{m,v}$. Note that $w \notin S_m$ and $v \notin S_m$. There exist w', v' such that $(w', v') \in J^h_{M,m}$ and $(x, y) \in \Omega_{m,w'} \times \Omega_{m,v'}$. Since $x \in \Omega_{m,w} \cap \Omega_{m,w'}$ and $y \in \Omega_{m,v} \cap \Omega_{m,v'}$, we see that $w' \in \Gamma_1(w)$, $v' \in \Gamma_1(v)$, $w' \notin S_m$ and $v' \notin S_m$. Therefore by (1),

$$\#(\{(x, y) | (x, y) \in J^h_{M,m}[\Omega], x \in \Omega_{m,v}, y \in \Omega_{m,w}\})$$

$$\leq \sum_{w' \in \Gamma_1(w), v' \in \Gamma_1(v)} \#(\Omega_{m,w'} \times \Omega_{m,v'}) \leq (L_0)^2 (L_*)^{2N+4}.$$

This shows the condition (N3) for $\overline{\Omega}^M$. The other conditions are straightforward and hence $\overline{\Omega}^M$ is a proper system of horizontal networks with indices $(M, (L_0)^2 (L_*)^{2N+4}, M, M)$.

Assume that $M \leq L_1$ for the moment. Let $(x, y) \in \Omega_{m,v} \times \Omega_{m,u}$ for some $u, v \in (T)_m$ with $u \in \Gamma_M(v)$. Since $M \leq L_1$, the condition (N5) implies that there exist $(x_1, \ldots, x_n)$ and $(w(1), \ldots, w(n))$ such that $w(i) \in \Gamma_{L_2}(u)$ for any $i = 1, \ldots, n$, $(x_i, x_{i+1}) \in E_m(w(i), w(i+1))$ for any $i = 1, \ldots, n-1$ and $x_1 = x, x_n = y, w(1) = u, w(n) = v$. Since $n - 1$ is no greater than the total number of edges in $\Gamma_{L_2}(u)$, we have

$$n - 1 \leq \#\left(\bigcup_{v_1, v_2 \in \Gamma_{L_2}(u)} E_m(v_1, v_2) \right) \leq (L_*)^{2L_2} L_0.$$

For any $f : \Omega_m \to \mathbb{R}$, by Lemma 4.6.17

$$|f(x) - f(y)|^p \leq C_h(p, n-1) \sum_{i=1}^{n-1} |f(x_i) - f(x_{i+1})|^p.$$

Let $(z_1, z_2) \in E_m$. Consider how many $(x, y) \in J^h_{M,m}$ there are such that (z_1, z_2) appears as (x_i, x_{i+1}) in the above inequality. We start with counting the number of possible u's. First there exist $\tau_1, \tau_2 \in (T)_m$ such that $z_1 \in \Omega_{m,\tau_1}$ and $z_2 \in \Omega_{\tau_2}$. Then $u \in \Gamma_{L_2}(\tau_1) \cap \Gamma_{L_2}(\tau_2)$. Therefore, possible number of $u's$ are at most

$$\#(\{\tau_1 | z_1 \in \Omega_{\tau_1, m}\}) \times \#(\{\tau_2 | z_2 \in \Omega_{\tau_2, m}\}) \times \#(\Gamma_{L_2}(\tau_1) \cap \Gamma_{L_2}(\tau_2)) \leq (L_*)^{L_2+2}.$$

For each u, we have $v \in \Gamma_M(u)$ and $(x, y) \in \Omega_{m,v} \times \Omega_{m,u}$. The possible number is

$$\#\left(\bigcup_{v \in \Gamma_M(u)} \Omega_{m,v} \times \Omega_{m,u} \right) \leq (L_*)^M (L_0)^2 (L_*)^{2N+2} = (L_0)^2 (L_*)^{M+2N+2}.$$

Combining these, we see that the possible number of (x, y) for which (z_1, z_2) appears as (x_i, x_{i+1}) is at most $(L_*)^{M+L_2+2N+4}(L_0)^2$, which is denoted by C_1. Then it follows that

$$\mathcal{E}_p(f | \Omega_m, J^h_{M,m}[\Omega]) \leq C_1 C_h(p, (L_*)^{2L_2} L_0) \mathcal{E}_p(f | \Omega_m, E_m).$$

So, we have finished the proof if $M \leq L$. For general situation, choosing n_0 so that $M \leq n_0 L_1$, we see that Ω is a proper system of horizontal networks with indices $(N, L_0, n_0 L_1, (n_0 - 1)L_1 + L_2)$. Thus replacing L_1 and L_2 by $n_0 L_1$ and $(n_0 - 1)L_1 + L_2$ respectively, we complete the proof for general cases.

(3) Set $m = |w| + k$. Note that if $v \in S_m$, then $\{(x, y) | (x, y) \in J^h_{M,m}[\Omega], x \in \Omega_{m,v}\} = \emptyset$. Therefore, no path in $\widetilde{C}_{w,k}(N_1, N_2, \overline{\Omega}^M)$ passes $\Omega_{m,v}$. Hence the value $f(v)$ has nothing to do with the criterion whether $f \in \mathcal{Q}_{w,k}(N_1, N_2, \overline{\Omega}^M)$ or not. Consequently, to get the infimum in the definition of $M_{p,k,w}(N_1, N_2, \overline{\Omega}^M)$, one may simply let $f(v) = 0$. $\square$

Lemma 4.9.3 (Sub-multiplicative Inequality) *Let* $\Omega = \{(\Omega_m, E_m)\}_{m \geq 0}$ *be a proper system of horizontal networks with indices* (N, L_0, L_1, L_2). *Then*

$$\mathcal{M}_{p,k+l}(0, M, \Omega^{(J)}_*) \leq C \mathcal{M}_{p,k}(0, M, \overline{\Omega}^{2M+J}) \mathcal{M}_{p,l}(0, M, \Omega^{(J)}_*),$$

for any $k, l, M, J \in \mathbb{N}$ *and* $p > 0$, *where* $C = L_* C_h(p, (L_*)^{N+1} L_0)$.

Similar sub-multiplicative inequalities for moduli of curve families have been shown in [10, Proposition 3.6] and [13, Lemma 3.8].

Proof Let $f \in \mathcal{Q}_{w,k}(0, M, \overline{\Omega}^{2M+J})$ and let $g_v \in \mathcal{Q}_{v,l}(0, M, \Omega^{(J)}_*)$ for any $v \in (T)_{|w|+k}$. Define $h : (T)_{|w|+k+l} \to [0, \infty)$ by

$$h(u) = \max\{f(x) g_v(u) | x \in \Omega_{|w|+k,v}, v \in \Gamma_M(\pi^l(u))\} \chi_{S^{k+l}(\Gamma_M(w))}(u).$$

Claim 1 $h \in \mathcal{Q}_{w,k+l}(0, M, \Omega^{(J)}_*)$.

Proof of Claim 1 Let $(u(1), \ldots, u(m)) \in \mathcal{C}_{w,k+l}(0, M, J)$. There exist $u(0) \in S^{k+l}(w)$ and $u(m+1) \in (T)_{|w|+k+l} \setminus S^{k+l}(\Gamma_M(w))$ such that $u(0) \in \Gamma_J(u(1))$ and $u(m+1) \in \Gamma_J(u(m))$. Set $v(i) = \pi^l(u(i))$ for $i = 0, \ldots, m+1$. Let $v_*(0) = v(0)$ and let $i_0 = 0$. Define n_*, $v_*(n)$ and i_n for $i = 1, \ldots, n_*$ inductively as follows: If

$$\max\{j | i_n \leq j \leq m, v(j) \in \Gamma_{2M}(v_*(n))\} = m,$$

then $n = n_*$. If

$$\max\{j | i_n \leq j \leq m, v(j) \in \Gamma_{2M}(v_*(n))\} < m,$$

then define

$$i_{n+1} = \max\{j | i_n \leq j \leq m, v(j) \in \Gamma_{2M}(v_*(n))\} + 1 \quad \text{and} \quad v_*(n+1) = v(i_{n+1}).$$

Since $v(i_{n+1} - 1) \in \Gamma_{2M}(v_*(n))$, it follows that $v_*(n + 1) \in \Gamma_{2M+J}(v_*(n))$. Hence $(v_*(1), \ldots, v_*(n_*)) \in \mathcal{C}_{w,k}(0, M, 2M + J)$. Moreover, since $\Gamma_M(v_*(n-1)) \cap \Gamma_M(v_*(n)) = \emptyset$ for $n = 1, \ldots, n_*$, there exists j_n such that $v(i_n - 1), \ldots, v(i_n - j_n) \in \Gamma_M(v_*(n))$ and $v(i_n - j_n - 1) \notin \Gamma_M(v_*(n))$. Then $(u(i_n - 1), u(i_n - 2), \ldots, u(i_n - j_n)) \in \mathcal{C}_{v_*(n),l}(0, M, J)$. Choose $x_i \in \Omega_{|w|+k,v_*(i)}$ for each $i = 1, \ldots, n_*$. Since $g_{v_*(n)} \in \mathcal{Q}_{v_*(n),l}(0, M, \Omega_*^{(J)})$, we have

$$\sum_{i=i_n-j_n}^{i_n-1} h(u(i)) \geq \sum_{i=i_n-j_n}^{i_n-1} f(x_n) g_{v_*(n)}(u(i)) \geq f(x_n).$$

This and the fact that $(x_1, \ldots, x_{n_*}) \in \widetilde{\mathcal{C}}_{k,w}(0, M, \overline{\Omega}^{2M+J})$ yield

$$\sum_{i=1}^{m} h(u(i)) \geq \sum_{j=1}^{n_*} f(x_j) \geq 1.$$

Thus Claim 1 has been verified. $\qquad\qquad\qquad\qquad\qquad\qquad\qquad\qquad\qquad\qquad\qquad\square$

Set $C_0 = C_h(p, (L_*)^{M+N+1} L_0)$. Then by Lemma 4.6.17,

$$h(u)^p \leq \left(\sum_{v \in \Gamma_M(\pi^l(u))} \sum_{x \in \Omega_{|w|+k,v}} f(x) g_v(u) \right)^p$$

$$\leq C_0 \sum_{v \in \Gamma_M(\pi^l(u))} \sum_{x \in \Omega_{|w|+k,v}} f(x)^p g_v(u)^p.$$

Set $M_{p,j,w'} = \mathrm{Mod}_p((T)_{|w'|+j}, J^h_{J,|w'|+j}, S^j(w'), (S^j(\Gamma_M(w')))^c)$. The above inequality and Claim 1 yield

$$M_{p,k+l,w} \leq \sum_{u \in (T)_{|w|+k+l}} h(u)^p \leq C_0 \sum_{v \in (T)_{|w|+k}} \sum_{x \in \Omega_{|w|+k,v}} \sum_{u \in (T)_{|w|+k+l}} f(x)^p g_v(u)^p.$$

Taking infimum regarding $g_v \in Q_{v,l}(0, M, \Omega_*^{(J)})$, and then supremum regarding $v \in (T)_{|w|+l}$, we have

$$M_{p,k+l,w} \leq C_0 \sum_{v \in (T)_{|w|+k}} \sum_{x \in \Omega_{|w|+k,v}} f(x)^p M_{p,l,v}$$

$$\leq C_0 \sum_{v \in (T)_{|w|+k}} \sum_{x \in \Omega_{|w|+k,v}} f(x)^p \mathcal{M}_{p,l}(0, M, \Omega_*^{(J)}).$$

Since $\sum_{v \in (T)_{|w|+k}} \sum_{x \in \Omega_{|w|+k,v}} f(x)^p \leq L_* \sum_{x \in \Omega_m} f(x)^p$, the above inequality leads to

$$M_{p,k+l,w} \leq C_0 L_* \mathcal{M}_{p,k}(0, M, \overline{\Omega}^{2M+J}) \mathcal{M}_{p,l}(0, M, \Omega_*^{(J)}).$$

Finally taking supremum regarding $w \in T$, we obtain the desired inequality

$$\mathcal{M}_{p,k+l}(0, M, \Omega_*^{(J)}) \leq C \mathcal{M}_{p,k}(0, M, \overline{\Omega}^{2M+J}) \mathcal{M}_{p,l}(0, M, \Omega_*^{(J)}),$$

where $C = C_0 L_*$. $\qquad\qquad\qquad\qquad\qquad\qquad\qquad\qquad\qquad\qquad\square$

Proof of Theorem 4.9.1 Write $\mathcal{M}_{p,j} = \mathcal{M}_{p,j}(0, M, \Omega_*^{(J)})$ and $\mathcal{M}'_{p,j} = \mathcal{M}_{p,j}(0, M, \overline{\Omega}^{2M+J})$. By Lemma 4.9.3,

$$\mathcal{M}_{p,mk+l} \leq (C\mathcal{M}'_{p,k})^m \mathcal{M}_{p,l}. \qquad\qquad (4.9.4)$$

Assume that $C\mathcal{M}'_{p,k} < 1 - \epsilon$ for some $\epsilon \in (0, 1)$. Then by Lemma 4.9.2-(3), for any $w \in T$, there exists $f_w \in Q_{w,k}(0, M, \overline{\Omega}^{2M+J})$ such that

$$C \sum_{x \in \Omega_{|w|+k}(S^k(\Gamma_M(w)) \setminus S_{|w|+k})} f_w(x)^p < 1 - \epsilon.$$

Since

$$\lim_{\delta \to 0} \max_{x \in [0,1]} (x^{p-\delta} - x^p) = 0,$$

there exists $\delta_* > 0$ such that $x^{p-\delta} \leq x^p + C_2^{-1}C^{-1}\epsilon/2$ for any $\delta \in (0, \delta_*]$ and $x \in [0, 1]$, where $C_2 = (N_*)^k(L_*)^{M+N+1}L_0$. By Lemma 4.9.2-(1),

$$\#(\Omega_{|w|+k}(S^k(\Gamma_M(w)))\backslash\mathcal{S}_{|w|+k}) \leq \#(S^k(\Gamma_M(w))) \max_{v \in S^k(\Gamma_M(w))\backslash\mathcal{S}_{|w|+k}} \#(\Omega_{|w|+k,v})$$

$$\leq (L_*)^M(N_*)^k(L_*)^{N+1}L_0 = C_2.$$

This and Lemma 4.9.2-(3) imply

$$C\mathcal{M}_{p-\delta,k,w}(0, M, \overline{\Omega}^{2M+J}) \leq C \sum_{x \in \Omega_{|w|+k}(S^k(\Gamma_M(w)))\backslash\mathcal{S}_{|w|+k}} f_w(v)^{p-\delta}$$

$$\leq C \sum_{x \in \Omega_{|w|+k}(S^k(\Gamma_M(w)))\backslash\mathcal{S}_{|w|+k}} f_w(x)^p + \frac{\epsilon}{2} \leq 1 - \frac{\epsilon}{2}.$$

Therefore $C\mathcal{M}'_{p-\delta,k} \leq 1 - \frac{\epsilon}{2}$. Replacing p in (4.9.4) by $p - \delta$ and taking $m \to \infty$, we see that $\underline{\mathcal{M}}_{p-\delta}(0, M, \Omega_*^{(J)}) = 0$. By Theorem 4.8.7, if $M \geq M_*$, it follows that $\dim_{AR}(X, d) \leq p - \delta < p$. Consequently, $C\mathcal{M}'_{p_*,k} \geq 1$ for any $k \geq 1$. Therefore, if $M \geq M_*$, then

$$C^{-1} \leq \underline{\mathcal{M}}_{p_*}(0, M, \overline{\Omega}^{2M+J}).$$

Using Lemma 4.8.6, we see that $0 < \underline{\mathcal{E}}_{p_*}(0, M, \overline{\Omega}^{2M+J})$. Then the inequality (4.9.3) shows that $0 < \underline{\mathcal{E}}_{p_*}(0, M, \Omega)$. Since $\underline{\mathcal{E}}_p(0, M, \Omega) \leq \underline{\mathcal{E}}_p(M', M, \Omega)$ for any $M' \in \{0, 1, \ldots, M - M_*\}$, we conclude that $0 < \underline{\mathcal{E}}_{p_*}(N_1, N_2, \Omega)$ for any $N_1, N_2 \geq 0$ with $N_2 \geq N_1 + M_*$. Again by Lemma 4.8.6, it follows that $0 < \underline{\mathcal{M}}_{p_*}(N_1, N_2, \Omega)$ as well. $\qquad\square$

Appendix A
Fact from Measure Theory

Proposition A.1 *Let $(X, \mathcal{M}, \mu)$ be measurable space and let $N \in \mathbb{N}$. If $U_i \in \mathcal{M}$ for any $i \in \mathbb{N}$ and*

$$\#(\{i \,|\, i \in \mathbb{N}, x \in U_i\}) \leq N \tag{A.1}$$

for any $x \in X$, then

$$\sum_{i=1}^{\infty} \mu(U_i) \leq N\mu\left(\bigcup_{i \in \mathbb{N}} U_i\right).$$

Proof Set $U = \cup_{i \in \mathbb{N}} U_i$. Define $U_{i_1 \ldots i_m} = \cap_{j=1,\ldots,m} U_{i_j}$. By (A.1), if $m > N$, then $U_{i_1 \ldots i_m} = \emptyset$. Fix $m \geq 0$ and let rearrange $\{U_{i_1 \ldots i_m} \,|\, i_1 < i_2 < \ldots < i_m\}$ so that

$$\{Y_j^m\}_{j \in \mathbb{N}} = \{U_{i_1 \ldots i_m} \,|\, i_1 < i_2 < \ldots < i_m\}.$$

Define

$$X_j^m = Y_j^m \backslash \left(\bigcup_{i \in \mathbb{N}, i \neq j} Y_i^m\right).$$

Then

$$U = \bigcup_{m=0}^{N} \left(\bigcup_{j \in \mathbb{N}} X_j^m\right)$$

J. Kigami, *Geometry and Analysis of Metric Spaces via Weighted Partitions*, Lecture Notes in Mathematics 2265, https://doi.org/10.1007/978-3-030-54154-5

and $X_j^m \cap X_l^k = \emptyset$ if $(m, j) \neq (k, l)$. This implies

$$\mu(U) = \sum_{m=0}^{N} \sum_{j \in \mathbb{N}} \mu(X_j^m).$$

Set $I_j = \{(k, l) | U_j \supseteq X_l^k \neq \emptyset\}$. Then by (A.1), we have $\#(\{j | (k, l) \in I_j\}) \leq N$ for any (k, l). This implies

$$\sum_{j=1}^{\infty} \mu(U_i) \leq N \sum_{m=0}^{N} \sum_{j \in \mathbb{N}} \mu(X_j^m) = N\mu(U). \qquad \square$$

Appendix B
List of Definitions, Notations and Conditions

Definitions

- adapted—Definition 2.4.1, Definition 2.4.7
- adjacent matrix—Definition 2.1.1
- Ahlfors regular—Definition 3.1.18
- Ahlfors regular conformal dimension—Definition 4.6.1
- Ahlfors regular metric—Definition 4.2.1
- balanced—Definition 4.1.2
- bi-Lipschitz (metrics)—Definition 3.1.9
- bi-Lipschitz (weight functions)—Definition 3.1.1
- bridge—Definition 2.5.3
- chain—Definition 2.2.1
- doubling (geometrically doubling)—Definition 2.2.5
- doubling (volume doubling)—Definition 3.3.3
- resolution—Definition 2.2.11
- degree of distortion—Definition 3.4.3
- end of a tree—Definition 2.1.2
- exponential—Definition 3.1.15
- (super-, sub-)exponential for metrics—Definition 3.6.2
- gentle—Definition 3.3.1
- geodesic—Definition 2.1.1
- geodesic ray—Definition 2.2.11
- Gromov product—Definition 2.5.6
- height (of a bridge)—Definition 2.5.3
- horizontal edge—Definition 2.2.11
- horizontal M-chain—Definition 4.1.1
- horizontally minimal—Definition 2.5.3
- hyperbolic—Definition 2.5.6

J. Kigami, *Geometry and Analysis of Metric Spaces via Weighted Partitions*, Lecture Notes in Mathematics 2265, https://doi.org/10.1007/978-3-030-54154-5

- hyperbolicity of a weight function –Definition 2.5.11
- infinite binary tree—Example 2.1.3
- infinite geodesic ray—Definition 2.1.2
- jpath—Definition 4.1.1
- jumping path—Definition 4.1.1
- locally finite—Definition 2.1.1
- minimal—Definition 2.2.1
- modulus—Definition 4.8.1
- m-separated—Definition 2.4.10
- neighborhood(graph)—Definition 2.1.1
- open set condition –Example 2.5.19
- partition—Definition 2.2.1
- path—Definition 2.1.1
- proper system of horizontal networks—Definition 4.6.5
- p-modulus of curves—Definition 4.8.1
- p-spectral dimension—Definition 4.7.7
- quasisymmetry—Definition 3.6.1
- rearranged resolution—Definition 2.5.10
- simple path—Definition 2.1.1
- strongly finite—Definition 2.2.4
- sub-exponential—Definition 3.1.15
- super-exponential—Definition 3.1.15
- thick—Definition 3.1.19
- tight—Definition 3.1.5
- tree—Definition 2.1.1
- tree with a reference point—Definition 2.1.2
- uniformly finite—Definition 3.1.15
- uniformly perfect—Definition 3.6.3
- visual pre-metric—Definition 2.3.8
- vertical edge—Definition 2.2.11
- volume doubling property w. r. t. a metric—Definition 3.3.3
- volume doubling property w. r. t. a weight function—Definition 3.3.5
- weakly M-adapted—Definition 2.5.14
- weight function—Definition 2.3.1

Notations

- B_w—Definition 2.2.1
- $\widetilde{B}_r^d(x, cr)$—Definition 2.5.14
- $\mathcal{B}$—Definition 2.2.11
- $\mathcal{B}_{\widetilde{T}^g, r}$—Definition 2.5.10
- $C_h(p, n)$—Lemma 4.6.17
- $\mathcal{C}(X, \mathcal{O}), \mathcal{C}(X)$—Definition 2.2.1

- $\mathcal{CH}_K(A, B)$—Definition 2.2.1
- $\mathcal{C}(V, E, U_1, U_2)$—Definition 4.8.1
- $\mathcal{C}_w^M$—Definition 4.1.2
- $\mathcal{C}_{w,k}(N_1, N_2, N)$—Definition 4.4.6
- $\widetilde{\mathcal{C}}_{w,k}(M_1, M_2, \Omega)$—(4.9.2)
- $\overline{d}_p^S(N_1, N_2, \Omega)$, $\underline{d}_p^S(N_1, N_2, \Omega)$—Definition 4.7.7
- $d_{(T,\mathcal{B})}$—Definition 2.2.11
- $D_M^g(x, y)$—Definition 2.4.3
- $D^g(x, y)$—Definition 2.4.3
- $\mathcal{D}(X, \mathcal{O})$—Definition 2.3.4
- $\mathcal{D}_A(X, \mathcal{O})$—Definition 3.1.9
- $\mathcal{D}_{A,e}(X, \mathcal{O})$—Definition 3.6.5
- $E_{g,r}^h$—Definition 2.5.10
- E_m^h, E^h: horizontal vertices—Definition 2.2.11
- $E_m(u, v)$—Definition 4.6.5
- $\mathcal{E}_p(f|V, E)$, $\mathcal{E}_p(V, E, V_1, V_2)$—Definition 4.6.2
- $\mathcal{E}_{p,k}(N_1, N_2, N)$, $\overline{\mathcal{E}}_p(N_1, N_2, N)$, $\underline{\mathcal{E}}_p(N_1, N_2, N)$—Definition 4.6.3
- $\mathcal{E}_{p,k,w}(N_1, N_2, \Omega)$, $\mathcal{E}_{p,k}(N_1, N_2, \Omega)$—Definition 4.6.8
- $\overline{\mathcal{E}}_p(N_1, N_2, \Omega)$, $\underline{\mathcal{E}}_p(N_1, N_2, \Omega)$—Definition 4.6.8
- $F(f)(x)$—Definition 4.8.2
- $\mathcal{F}_F(V, V_1, V_2)$—Definition 4.6.2
- $\mathcal{F}_M(V, E, U_1, U_2)$—Definition 4.8.1
- g_d, g_μ—Definition 2.3.4
- $G(g)(x)$—Definition 4.8.2
- $\mathcal{G}(T)$—Definition 2.3.1
- $\mathcal{G}_e(T)$—Definition 3.5.1
- h_*—Definition 3.2.3
- h_r—Corollary 2.5.13
- $I_{\mathcal{E}}(N_1, N_2, N)$—Definition 4.6.3
- $\overline{I}_{\mathcal{E}}(N_1, N_2, \Omega)$, $\underline{I}_{\mathcal{E}}(N_1, N_2, \Omega)$—Definition 4.6.8
- $\overline{I}_{\mathcal{M}}(N_1, N_2, \Omega)$, $\underline{I}_{\mathcal{M}}(N_1, N_2, \Omega)$—Definition 4.8.5
- $J_{M,n}^h(K)$, $J_M^h(K)$, $J_M^v(K)$, $J_M(K)$—Definition 4.1.1
- $J_{M,n}^h[\Omega]$—Lemma 4.9.2
- $K^{(q)}$—Definition 4.4.5
- K_w—Definition 2.2.1
- $\ell_M^\varphi(\mathbf{p})$—Definition 4.1.8
- L_*—Definition 4.3.2
- $L_g(\mathbf{p})$—Definition 4.1.8
- $\mathrm{Mod}_p(V, E, U_1, U_2)$—Definition 4.8.1
- $\mathcal{M}_{p,k,w}(N_1, N_2, \Omega)$, $\mathcal{M}_{p,k}(N_1, N_2, \Omega)$—Definition 4.8.5
- $\underline{\mathcal{M}}_p(N_1, N_2, \Omega)$, $\overline{\mathcal{M}}_p(N_1, N_2, \Omega)$—Definition 4.8.5
- $\mathcal{M}_P(X, \mathcal{O})$—Definition 2.3.4
- N_*—Definition 4.3.2
- $\overline{N}_*$, $\underline{N}_*$—Definition 4.6.11

- $N_g(w)$—Definition 3.5.3
- O_w—Definition 2.2.1
- $\mathcal{P}(V, E)$—Definition 4.8.1
- $\mathcal{Q}_{w,k}(M_1, M_2, \Omega)$—(4.9.1)
- $\overline{R}_p(N_1, N_2, \Omega), \underline{R}_p(N_1, N_2, \Omega)$—Definition 4.7.1
- $\mathcal{R}_\kappa^0, \mathcal{R}_\kappa^1$—Definition 3.4.3
- $S^m(A)$—Definition 3.5.3
- $S(\cdot)$—Definition 2.1.2
- $\mathcal{S}_m$—Lemma 4.9.2
- $(T)_m$—Definition 2.1.2
- $T^{(N)}, (T^{(N)}, \mathcal{A}^{(N)}, \phi), T_m^{(N)}$—Example 2.1.3
- $T^{(q)}$—Definition 4.4.5
- T_w—Definition 2.1.6
- $(T, \mathcal{B})$—Definition 2.2.11
- $\widetilde{T}^{g,r}, (\widetilde{T}^{g,r}, \mathcal{B}_{\widetilde{T}^{g,r}})$—Definition 2.5.10
- $U_M^g(x, s)$—Definition 2.3.6
- $U_M(w, K)$—Definition 4.1.1
- $\Gamma_M(w, K)$—Definition 4.1.1
- $\delta_M^g(x, y)$—Definition 2.3.8
- $\kappa(\cdot)$—Definition 3.4.3
- Λ_s^g—Definition 2.3.1
- $\Lambda_{s,M}^g(\cdot)$—Definition 2.3.6
- $\Omega^k(w, n)$—(4.6.2)
- $\Omega^{k,c}(w, n)$—(4.6.3)
- $\Omega_m(U)$—(4.6.1)
- $\Omega_{m,w}$—Definition 4.6.5
- $\Omega_*^{(N)}$—Example 4.6.6
- $\overline{\Omega}^M$—Lemma 4.9.2
- π—Definition 2.1.2
- $\pi^{(T,\mathcal{A},\phi)}$—Remark after Definition 2.1.2
- $\pi_g^*(w)$—Definition 3.5.3
- $\Pi_M^\varphi(w)$—Definition 4.4.1
- $\Pi_M^{g,k}(w)$—Definition 4.4.6
- ρ_*—Definition 2.1.6
- Σ: the collection of ends—Definition 2.1.2
- Σ^w, Σ_v^w—Definition 2.1.2
- Σ and Σ_v; abbreviation of Σ^ϕ and Σ_v^ϕ respectively,
- $\Sigma^{(N)}$—Example 2.1.5
- $\#(\cdot)$—Eq. (1.2.3)
- $|w, v|$—Definition 3.5.3
- $\overline{wv}$: the geodesic between w and v of a tree,—Definition 2.1.1
- $(w|v)_{((T,\mathcal{B}),\phi)}, (w|v)$: Gromov product—Definition 2.5.6
- $|w|$—Definition 2.1.2

- $|w|_{(T,\mathcal{A},\phi)}$—Remark after Definition 2.1.2
- $w \wedge v$—Definition 2.1.6
- $[\omega]_m$—Definition 2.1.6
- $\langle \cdot \rangle_M$—Definition 4.4.1

Equivalence Relations

- $\underset{AC}{\sim}$—Definition 3.1.3
- $\underset{BL}{\sim}$ relation on weight functions—Definition 3.1.1
- $\underset{BL}{\sim}$ relation on metrics—Definition 3.1.9
- $\underset{GE}{\sim}$—Definition 3.3.1
- $\underset{QS}{\sim}$—Definition 3.6.1

Conditions

- (ADa), (ADb)$_M$—Theorem 2.4.5
- (BF1), (BF2)—Sect. 4.3
- (BL), (BL1), (BL2), (BL3)—Theorem 3.1.8
- (EV)$_M$, (EV2)$_M$, (EV3)$_M$, (EV4)$_M$, (EV5)$_M$—Theorem 2.4.12
- (G1), (G2), (G3)—Definition 2.3.1
- (N1), (N2), (N3), (N4), (N5)—Definition 4.6.5
- (P1), (P2)—Definition 2.2.1
- (SQ1), (SQ2), (SQ3)—Sect. 3.4
- (SF)—Strongly finite, (1.2.3)
- (TH)—Few lines before Definition 1.2.1
- (TH1), (TH2), (TH3), (TH4)—Theorem 3.2.3
- (VD1), (VD2), (VD3), (VD4)—Theorem 3.3.9

Bibliography

1. R. Adler, Symbolic dynamics and Markov partitions. Bull. Amer. Math. Soc. **35**, 1–56 (1998)
2. M.T. Barlow, *Diffusion on Fractals*. Lecture Notes in Mathematics, vol. 1690 (Springer, Berlin, 1998)
3. M.T. Barlow, R.F. Bass, The construction of Brownian motion on the Sierpinski carpet. Ann. Inst. Henri Poincaré **25**, 225–257 (1989)
4. M.T. Barlow, R.F. Bass, Brownian motion and harmonic analysis on Sierpinski carpets. Canad. J. Math. **51**, 673–744 (1999)
5. M.T. Barlow, E.A. Perkins, Brownian motion on the Sierpinski gasket. Probab. Theory Related Fields **79**, 542–624 (1988)
6. M.T. Barlow, R.F. Bass, J.D. Sherwood, Resistance and spectral dimension of Sierpinski carpets. J. Phys. A Math. Gen. **23**, L253–L258 (1990)
7. M. Bonk, B. Kleiner, Conformal dimension and Gromov hyperbolic groups with 2-sphere boundary. Geom. Topol. **9**, 219–246 (2005)
8. M. Bonk, D. Meyer, *Expanding Thurston Maps*. Mathematical Surveys and Monographs (American Mathematical Society, Providence, 2017)
9. M. Bonk, E. Saksman, Sobolev spaces and hyperbolic fillings. J. Reine Angew. Math. **737**, 161–187 (2018)
10. M. Bourdon, B. Kleiner, Combinatorial modulus, the combinatorial Loewner property, and Coxeter groups. Group Geom. Dyn. **7**, 39–107 (2013)
11. M. Bourdon, H. Pajot, Cohomologie ℓ_p et espaces de Besov. J. Reine Angew. Math. **558**, 85–108 (2003)
12. S. Buyalo, V Schroeder, *Elements of Asymptotic Geometry*. EMS Monographs in Mathematics (European Mathematical Society, Zürich, 2007)
13. M. Carrasco Piaggio, On the conformal gauge of a compact metric space. Ann. Sci. Ecole. Norm. Sup. **46**, 495–548 (2013)
14. M. Christ, A T(b) theorem with remarks on analytic capacity and the Cauchy integral. Colloq. Math. **60/61**, 601–628 (1990)
15. G. Elek, The ℓ^p-cohomology and the conformal dimension of hyperbolic cones. Geom. Ded. **68**, 263–279 (1997)
16. J. Heinonen, *Lectures on Analysis on Metric Spaces* (Springer, Berlin, 2001)
17. T. Hytönen, A. Kairema, Systems of dyadic cubes in a doubling metric space. Colloq. Math. **126**, 1–33 (2012)

© The Editor(s) (if applicable) and The Author(s), under exclusive license
to Springer Nature Switzerland AG 2020
J. Kigami, *Geometry and Analysis of Metric Spaces via Weighted Partitions*,
Lecture Notes in Mathematics 2265, https://doi.org/10.1007/978-3-030-54154-5

18. I. Kapovich, N. Benakli, Boundaries of hyperbolic groups, in *Combinatorial and Geometric Group Theory*. Contemporary Mathematics, vol. 296 (American Mathematical Society, Providence, 2002), pp. 39–93
19. J. Kigami, *Analysis on Fractals*. Cambridge Tracts in Mathematics, vol. 143 (Cambridge University Press, Cambridge, 2001)
20. J. Kigami, *Volume Doubling Measures and Heat Kernel Estimates on Self-similar Sets*. Memoirs of the American Mathematical Society, vol. 199 (American Mathematical Society, Providence, 2009), p. 932
21. J. Kigami, *Resistance Forms, Quasisymmetric Maps and Heat Kernel Estimates*. Memoirs of the American Mathematical Society, vol. 216 (American Mathematical Society, Providence, 2012), p. 1015
22. J. Kigami, Quasisymmetric modification of metrics on self-similar sets, in *Geometry and Analysis of Fractals*, ed. by D.-J. Feng, K.-S. Lau. Springer Proceedings in Mathematics & Statistics, vol. 88 (Springer, Berlin, 2014), pp. 253–282
23. J. Kigami, M.L. Lapidus, Weyl's problem for the spectral distribution of Laplacians on p.c.f. self-similar fractals. Comm. Math. Phys. **158**, 93–125 (1993)
24. S. Kusuoka, X.Y. Zhou, Dirichlet forms on fractals: Poincaré constant and resistance. Probab. Theory Related Fields **93**, 169–196 (1992)
25. K.-S. Lau, X.-Y. Wang, Self-similar sets as hyperbolic boundaries. Indiana Univ. Math. J. **58**, 1777–1795 (2009)
26. J. Lindquist, Weak capacity and critical exponents, Preprint
27. J.M. Mackay, J.T. Tyson, *Conformal Dimension, Theory and Application*. University Lecture Series, vol. 54 (American Mathematical Society, Providence, 2010)
28. R. Shimizu, Parabolic index of an infinite graph and Ahlfors regular conformal dimension of a self-similar set, preprint
29. J. Väisälä, Gromov hyperbolic spaces. Expo. Math. **23**, 187–231 (2005)

LECTURE NOTES IN MATHEMATICS 🐎 Springer

Editors in Chief: J.-M. Morel, B. Teissier;

Editorial Policy

1. Lecture Notes aim to report new developments in all areas of mathematics and their applications – quickly, informally and at a high level. Mathematical texts analysing new developments in modelling and numerical simulation are welcome.

   Manuscripts should be reasonably self-contained and rounded off. Thus they may, and often will, present not only results of the author but also related work by other people. They may be based on specialised lecture courses. Furthermore, the manuscripts should provide sufficient motivation, examples and applications. This clearly distinguishes Lecture Notes from journal articles or technical reports which normally are very concise. Articles intended for a journal but too long to be accepted by most journals, usually do not have this "lecture notes" character. For similar reasons it is unusual for doctoral theses to be accepted for the Lecture Notes series, though habilitation theses may be appropriate.

2. Besides monographs, multi-author manuscripts resulting from SUMMER SCHOOLS or similar INTENSIVE COURSES are welcome, provided their objective was held to present an active mathematical topic to an audience at the beginning or intermediate graduate level (a list of participants should be provided).

   The resulting manuscript should not be just a collection of course notes, but should require advance planning and coordination among the main lecturers. The subject matter should dictate the structure of the book. This structure should be motivated and explained in a scientific introduction, and the notation, references, index and formulation of results should be, if possible, unified by the editors. Each contribution should have an abstract and an introduction referring to the other contributions. In other words, more preparatory work must go into a multi-authored volume than simply assembling a disparate collection of papers, communicated at the event.

3. Manuscripts should be submitted either online at www.editorialmanager.com/lnm to Springer's mathematics editorial in Heidelberg, or electronically to one of the series editors. Authors should be aware that incomplete or insufficiently close-to-final manuscripts almost always result in longer refereeing times and nevertheless unclear referees' recommendations, making further refereeing of a final draft necessary. The strict minimum amount of material that will be considered should include a detailed outline describing the planned contents of each chapter, a bibliography and several sample chapters. Parallel submission of a manuscript to another publisher while under consideration for LNM is not acceptable and can lead to rejection.

4. In general, **monographs** will be sent out to at least 2 external referees for evaluation.

   A final decision to publish can be made only on the basis of the complete manuscript, however a refereeing process leading to a preliminary decision can be based on a pre-final or incomplete manuscript.

   Volume Editors of **multi-author works** are expected to arrange for the refereeing, to the usual scientific standards, of the individual contributions. If the resulting reports can be forwarded to the LNM Editorial Board, this is very helpful. If no reports are forwarded or if other questions remain unclear in respect of homogeneity etc, the series editors may wish to consult external referees for an overall evaluation of the volume.

5. Manuscripts should in general be submitted in English. Final manuscripts should contain at least 100 pages of mathematical text and should always include

   – a table of contents;
   – an informative introduction, with adequate motivation and perhaps some historical remarks: it should be accessible to a reader not intimately familiar with the topic treated;
   – a subject index: as a rule this is genuinely helpful for the reader.
   – For evaluation purposes, manuscripts should be submitted as pdf files.

6. Careful preparation of the manuscripts will help keep production time short besides ensuring satisfactory appearance of the finished book in print and online. After acceptance of the manuscript authors will be asked to prepare the final LaTeX source files (see LaTeX templates online: https://www.springer.com/gb/authors-editors/book-authors-editors/manuscriptpreparation/5636) plus the corresponding pdf- or zipped ps-file. The LaTeX source files are essential for producing the full-text online version of the book, see http://link.springer.com/bookseries/304 for the existing online volumes of LNM). The technical production of a Lecture Notes volume takes approximately 12 weeks. Additional instructions, if necessary, are available on request from lnm@springer.com.

7. Authors receive a total of 30 free copies of their volume and free access to their book on SpringerLink, but no royalties. They are entitled to a discount of 33.3 % on the price of Springer books purchased for their personal use, if ordering directly from Springer.

8. Commitment to publish is made by a *Publishing Agreement*; contributing authors of multiauthor books are requested to sign a *Consent to Publish form*. Springer-Verlag registers the copyright for each volume. Authors are free to reuse material contained in their LNM volumes in later publications: a brief written (or e-mail) request for formal permission is sufficient.

Addresses:
Professor Jean-Michel Morel, CMLA, École Normale Supérieure de Cachan, France
E-mail: moreljeanmichel@gmail.com

Professor Bernard Teissier, Equipe Géométrie et Dynamique,
Institut de Mathématiques de Jussieu – Paris Rive Gauche, Paris, France
E-mail: bernard.teissier@imj-prg.fr

Springer: Ute McCrory, Mathematics, Heidelberg, Germany,
E-mail: lnm@springer.com

Printed in the United States
By Bookmasters